Palgrave Studies in Nineteenth-Century Writing and Culture

Series Editor
Joseph Bristow
Department of English
University of California, Los Angeles
Los Angeles, CA, USA

Palgrave Studies in Nineteenth-Century Writing and Culture is a monograph series that aims to represent innovative and interdisciplinary research on literary and cultural works that were produced from the time of the Napoleonic Wars to the fin de siècle. Attentive to the historical continuities between 'Romantic' and 'Victorian', the series features studies that assist in reassessing the meaning of these terms during a century marked by diverse cultural, literary, and political movements. The aim of the series is to look at the increasing influence of different types of historicism on our understanding of literary forms and genres. It reflects a broad shift from critical theory to cultural history that has affected not only the 1800-1900 period but also every field within the discipline of English literature. All titles in the series seek to offer fresh critical perspectives and challenging readings of both canonical and non-canonical writings of this era.

More information about this series at
http://www.palgrave.com/gp/series/14607

Ryan Sweet

Prosthetic Body Parts in Nineteenth-Century Literature and Culture

Wellcome Trust

palgrave
macmillan

Ryan Sweet
Swansea University
Swansea, West Glamorgan, UK

ISSN 2634-6494 ISSN 2634-6508 (electronic)
Palgrave Studies in Nineteenth-Century Writing and Culture
ISBN 978-3-030-78588-8 ISBN 978-3-030-78589-5 (eBook)
https://doi.org/10.1007/978-3-030-78589-5

Cover illustration: Classic Image / Alamy Stock Photo
Cover design: eStudioCalamar

This Palgrave Macmillan imprint is published by the registered company Springer Nature Switzerland AG.
The registered company address is: Gewerbestrasse 11, 6330 Cham, Switzerland

This book is dedicated to the prosthesis-using women who have shaped my life: Mum (hip), Nanny Sweet (two knees and a pacemaker), and Jelly Nan (hip).

Acknowledgements

Professional and Personal

This book is the result of much collaborative effort. Without the various generous organisations and individuals who have supported me and my research, this work would not be appearing before you. *Prosthetic Body Parts in Nineteenth-Century Literature and Culture* is developed from the research that I completed for my PhD at the University of Exeter from 2012 to 2016. I was fortunate to be funded by an Arts and Humanities Research Council Medical Humanities Studentship. The revision of my thesis into a book took place in large part while I was a Wellcome Trust Ins titutional Strategic Support Fund Fellow at the University of Leeds from 2017 to 2018. I am very grateful for the generous financial support of the AHRC and the Wellcome Trust. The Wellcome Trust also funded the open-access publication of this book. This funding is testament to their support for early career researchers and their commitment to research accessibility. It would be remiss not to acknowledge the other funding bodies whose support facilitated discussions of earlier versions of the material found here across the United Kingdom and overseas. Thanks to the British Society for Literature and Science, the Social Sciences and Humanities Research Council of Canada, the Dickens Project (University of California, Santa Cruz), and the College of Humanities at the University of Exeter. I am also hugely thankful to the departments and institutions that put faith in me to prepare and deliver this book: the Department of English, Centre for Victorian Studies, and Centre for Medical History at Exeter; the School of English, Centre for Medical Humanities, and the

Humanities Research Institute at Leeds; the School of Humanities and Performing Arts at the University of Plymouth; and the Faculty of Humanities and Social Sciences at Swansea University.

There are many individuals who deserve acknowledgement. For a start, I would like to thank my PhD supervisors, Jason Hall and Richard Noakes, for their hard work and generous support before, during, and after my PhD. Their guidance and astute comments helped to shape this book. Perhaps more so than anyone else, I owe Professor Hall a huge debt of gratitude. If not for his encouragement, I would not have pursued a postgraduate degree in the first place. I already look forward to working with Professor Hall on future projects. Angelique Richardson and Graeme Gooday made the examination of my PhD thesis intellectually stimulating, rewarding, and enjoyable. They have been excellent mentors since the viva too. I have been blessed to have the support of many sagacious colleagues, whose guidance has shaped me into the academic that I am today. Vike Plock, Stuart Murray, and Mark Jackson, and Kyriaki Hadjiafxendi and Mandy Bloomfield, all deserve special mention alongside my supervisors and examiners. Preparing this typescript was also very formative. Thank you to Joseph Bristow, the anonymous reviewers, and the editorial team at Palgrave Macmillan for making me a better scholar.

I would also like to thank those who contributed informal feedback on this work. There are many who asked stimulating questions or provided thoughtful comments at talks, conferences, and other events—far too many to list entirely—but I would like to directly thank Paul Williams, Corinna Wagner, Andy Brown, Matt Hayler, Vanessa Warne, Karen Bourrier, Jennifer Esmail, Keir Waddington, Martin Willis, Sue Zemka, James Mussell, Richard Salmon, Emma Curry, Helen Goodman, Regenia Gagnier, John Plunkett, Heather Tilley, Tricia Zakreski, Joe Kember, Claire Jones, Neil Pemberton, Andrew Mangham, Ruth Heholt, Clare Stainthorp, Michelle Webb, and Joanne Parsons. To the many support staff and librarians—particularly those at Exeter, Leeds, and the Wellcome Library—whose tireless help and troubleshooting facilitated my research, please accept my sincere thanks. Cathryn Baker, Morwenna Hussey, Claire Keyte, and Pamela Rhodes, all deserve special mention.

The completion of this book owes much to the encouragement of a wide network of academic friends, many of whom I mention above. As someone from a working-class Cornish background, I was initially concerned about how welcome I would feel within the academy, but this anxiety was misplaced. I would like to thank all my close colleagues (past and present) from Exeter, Bath Spa, Plymouth, Leeds, and Swansea. Special thanks to my wonderful PhD peers, including Wei Hsien-Wan, Richard Graham, Zoe Bullaitis, Sarah Jones, Simon Peplow, Lorna Peplow, Ed Taylor, Mike Rose-Steel, Esther van Raamsdonk, Tom Chadwick, Eddie Falvey, Fred Cooper, Isabel Galleymore, Phil Child, Georgina Hunter, Jonathan Memel, Sophie David, Anna-Marie Linnell, Sarah Daw, Ryan Patterson, Tara Etherington, James Parker, Joe Hickinbottom, George Twigg, and Jasmine Hunter-Evans—we made it together! To these good folks and everyone else who has brought happiness to my early career in academia, I am eternally grateful.

Many thanks must also go to my parents. Though what we do for work is very different, you have been nothing but encouraging and supportive. Your faith in me has inspired perseverance. For the same reasons, I must also thank my brothers (Glen and Marc), Nanny Sweet, my extended family, and my future in-laws, Jane and Martin. For providing much-needed distraction, companionship, and beer drinking, I would like to raise a glass to my many other non-academic friends. Beadle, Pidge, Pat, Gary, Tim, Chris, Lorie, Jessie, Anya, Sophie, Paul, and Neil: the next drink is on me!

Finally, I would like to thank Jen for her constant love, support, and patience. I know that you do not "get" Victorian studies—using my copy of *The Way We Live Now* as a coffee coaster was a giveaway—but I could not have done this without you.

Reuse of Copyrighted Material

Contents

List of Figures

CHAPTER 1

Introduction

How were artificial limbs, eyes, teeth, and hair imagined and presented in nineteenth- and early twentieth-century literature, journalism, and visual culture? In what ways did cultural imaginaries of prostheses reflect or respond to real-life developments of these technologies and the lived realities of their users? To what extent did these sources endorse or challenge the social mandate for physical normalcy that fed the appetite for and development of prosthetic devices that could conceal physical difference from public view? And how were portrayals of prostheses inflected by social inequalities related to social class, gender, and age? These questions provide the stimulus for the study that follows. Such lines of enquiry matter if we are to better understand where the enduring hegemony of physical wholeness comes from and how society responded to this concept when it emerged most strongly. Responding to these questions also helps us to comprehend how normalcy became entwined with and reinforced by other social prejudices and how ascendant cultural forms such as literature, media sources, and visual artwork played vital roles in challenging normative thinking. By revisiting these materials, we learn how we might build on this approach today in order to develop a less stigmatizing social system.

Prosthetic Body Parts in Nineteenth-Century Literature and Culture takes as its source materials British and American literary writings, print media, and visual artworks from the nineteenth and early twentieth century. These works create a prosthesis consciousness—that is to say, an imaginative focus on the extent to which prostheses successfully substitute

R. Sweet, *Prosthetic Body Parts in Nineteenth-Century Literature and Culture*, Palgrave Studies in Nineteenth-Century Writing and Culture, https://doi.org/10.1007/978-3-030-78589-5_1

for lost body parts. They also reorient our understanding of the period's attitudes to concepts of agency, normalcy, and difference. In terms of canonical literature, I analyse many of the best-remembered fictional prosthesis users, including Captain Ahab from Herman Melville's *Moby-Dick* (1851), Brevet Brigadier General John A. B. C. Smith from Edgar Allan Poe's "The Man That Was Used Up" (1839), and Captain Cuttle and Silas Wegg from Charles Dickens's *Dombey and Son* (1846–1848) and *Our Mutual Friend* (1864–1865), respectively.[1] Alongside these familiar fictional prosthesis users, I investigate works by other canonical authors, whose focus on prostheses has gone under the radar, including Wilkie Collins's *Armadale* (1866), *The Law and the Lady* (1875), and *The Black Robe* (1881); Thomas Hardy's *The Woodlanders* (1886–1887) and *Jude the Obscure* (1895); H. Rider Haggard's *King Solomon's Mines* (1885); and Arthur Conan Doyle's *The Sign of Four* (1890).[2] Together with these well-known fictions, I explore works in verse and prose by less-well-remembered writers, such as Robert Williams Buchanan, Thomas Hood, and Henry Clay Lewis, as well as many unsigned sketches, short stories, and journalistic pieces that appeared in newspapers and magazines, ranging from weekly penny publications aimed at middle- and lower-class adolescent readerships, such as *Chums*, to more expensive monthly periodicals written for middle- and high-brow adult readers, such as *Tait's Edinburgh Magazine*. In addition to these sources, I examine visual materials from graphic magazines (e.g. *Fun*), advertisements, and fine art (including works by J. T. Smith, G. M. Woodward, and Louis Leopold Boilly). Moreover, I investigate the prosthetic body part in early short films, such as J. Stuart Blackton's *The Thieving Hand* (1908).[3] What draws these sources together is their centralization of the prosthetic part and their engagement with conceptualizations of physical wholeness.

Following in the footsteps of recent important studies of nineteenth-century physical difference and prostheses, such as Erin O'Connor's *Raw Material* (2000), Jennifer Esmail's *Reading Victorian Deafness* (2013), and Claire L. Jones's *Rethinking Modern Prosthesis* (2017), this study analyses sources from both sides of the Atlantic.[4] In the nineteenth century, the trade of prostheses was thoroughly transatlantic. Successful artificial limb makers of the American North, such as B. Frank Palmer and A. A. Marks—who benefited from being approved suppliers for the US government's scheme to provide its maimed Civil War veterans with artificial legs—successfully marketed their devices to British clients. Meanwhile, British limb maker Frederick Gray supplied artificial legs to Confederate officers during the 1860s.[5] During the transatlantic success of American artificial limbs even before the Civil War (1861–1865), as Gordon Phillips notes, Palmer legs were used by 1200 amputees in Britain.[6] Similarly, the implementation of crowns and bridges, which became more popular replacements for lost teeth than partial dentures from the 1870s onwards, became known as "American dentistry" in Britain, reflecting the superiority of American dental expertise in the second half of the century. Figure 1.1, which shows an 1890s advertisement for Mr. Foley's artificial teeth and dentistry, underscores how the adjective "American" (which is centred in enlarged, emboldened, and accentuated font) was used to confer quality and authority. Conversely, British writers, such as Dickens, were admired by and potentially inspired the works of American authors of prosthesis narratives, such as Poe.[7] British prosthesis narratives, such as the ballad "Cork Leg" (*c.*1830) and Thomas Hood's *Miss Kilmansegg and Her Precious Leg* (1840–1841),

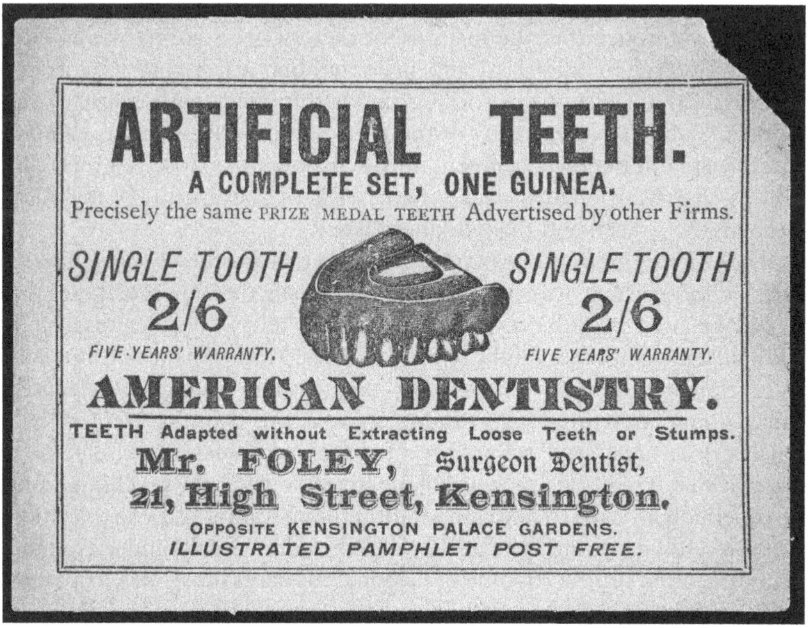

Fig. 1.1 A circa 1896 advertisement for Mr. Foley's artificial teeth and "American dentistry." "Artificial Teeth: A Complete Set, One Guinea," c. 1896, illustrated advertisement, Wellcome Collection, London. Image courtesy of the Wellcome Collection. https://wellcomecollection.org/works/j625k4vh.CCBY4.0

which drew into question aspects of artificial limb design, such as weight, showiness, and sophistication, were so popular and iconic that they were mentioned and sometimes even recited in the prosthesis catalogues of prominent American artificial limb makers, such as John S. Drake and A. A. Marks.[8] As these examples demonstrate, there existed a two-way dialogue across the Atlantic in terms of both the trade and the culture of prostheses in the nineteenth and early twentieth centuries.

My selection of sources is informed largely by two factors: first, the centrality of prostheses or prosthesis users; and secondly, the extent to which prosthesis narratives are characteristic of larger representational tropes. Besides the well-known works mentioned above, the chapters that follow engage with several largely forgotten novels, poems, short stories, jests, and comics, in which prosthetic body parts are the primary focus—such as André de Blaumont's short story "My Fiancé's Glass Eye" (1894), which tells a narrative of an engagement that is almost broken off after the bride-to-be is mistaken into thinking that her lover is a glass-eye user.[9] Readers will notice that the majority of these sources are by white, educated, Western, middle-class, and male authors, a fact that mirrors the authorial dominance of this social group within nineteenth-century print culture. Since, however, many of the users and prospective users of prostheses were not middle-class men, my discussion looks closely at representations that were pitched at a broad range of social groups, including women, the elderly, and the working classes.

There are significant distinctions between the prostheses that I discuss. First, of those listed, artificial limbs are potentially the only prostheses that would be used by subjects whom we might today consider disabled—though some amputees might reject this label. Most of us would not consider someone thought to be missing hair, teeth, or even an eye disabled. Still, I draw upon a disability studies approach to consider each of these devices. I certainly have no wish to homogenize physical difference or to suggest that conditions such as baldness, are somatically, psychologically, or experientially akin to limb amputation, but I want to expose that those perceived to be missing hair, teeth, or an eye in the period under examination were often subject to some of the same stigma as those with lost limbs.

Part of the prejudice faced by those who were perceived to be missing body parts stemmed from the social preference for physical wholeness: a predilection culminating from several factors, including the rise of bodily statistics, the vogue for physiognomy, and changing models of work. The other focus of discrimination centred on the use of artifice, a practice seen as dishonest, deceitful, and, at times, fraudulent. For example, writing about prosthesis manufacture for *Once a Week* in 1859, the author-physician Andrew Wynter lamented: "What member is there in this artful age that we can depend upon as genuine?" Wynter emphasized both the apparent scale of prosthesis use and the extent to which they could dupe dependencies on physical normalcy as a signifier of trustworthiness.[10] It cannot, however, be denied that those missing limbs faced greater stigma in certain regards. For instance, in "autobiographical" accounts such as John Brown's "A Memoir of Robert Blincoe, an Orphan Boy" (1832), *A Narrative of the Experience and Sufferings of William Dodd, a Factory Cripple* (1841), and Henry Mayhew's *London Labour and the London Poor* (1851), amputees experienced significant discrimination from potential employers.[11] By comparison, those individuals using what we might call primarily cosmetic prostheses, such as dentures, artificial eyes, and wigs, were arguably more susceptible to casual physical scrutiny, cruel jests, and accusations of fraudulence—as evidenced by the numerous unkind stories, cartoons, and jokes about users of these devices that appeared in contemporary newspapers and magazines.[12] The users of "cosmetic" prostheses, if discovered, were subject to scorn for duplicity since some believed that their use of such prostheses cheated popular methods of assessing character by looks. This view was made manifest in a comic article that appeared in *Pick-Me-Up* in 1892, which equated "[p]aint, powder, false teeth, false hare, and … a false buzzum" with "a false hart! [*sic*]."[13] Users of these kinds of prostheses were also often accused of vanity, a serious charge at the time.[14] In his 1851 *Household Words* article "Eyes Made to Order," William Blanchard Jerrold, for instance, explained: "To some persons a wig is the type of a false and hollow age; an emblem of deceit; a device of ingenious vanity, covering the wearer with gross and unpardonable deceit."[15]

The users of artificial limbs were also subject to a degree of the same stigma, especially if they were deemed to be concealing their impairments to better their social positions, as with the homeless prosthesis users described in the 1877 *All the Year Round* article "Mr. Wegg and His Class"—though they were generally treated with more sympathy.[16] The users of wigs, artificial eyes, and dentures, especially if single and female and/or elderly, were more regularly and directly mocked in public venues, including newspapers and magazines. In 1907, the *Penny Illustrated Paper* quoted a pastor from Liverpool who declared to his congregation that "a wig was a foolish relic of the bad old days, a thatching of one's roof by an artificial process, and one of the few foolish things women have never done."[17] Depictions of amputees, especially children and veterans, were sometimes tinged with sympathy, as in the case of Sir Hubert von Herkomer's famous 1875 oil painting *The Last Muster*, which depicts a group of Chelsea Pensioners, some of whom are wooden-leg users, at a Sunday service at the Royal Hospital Chelsea. But wig users, by contrast, were considered fair game for jests.[18] The following, for example, appeared in William Carew Hazlitt's *New London Jest Book* in 1871:

> Walking one day, to dine with a friend, some miles from Cambridge, Dr Parr was overtaken by a heavy fall of rain, and not being able to procure shelter, was completely drenched before he reached his destination. With linen and clothes his friend was able to furnish him, but his handkerchief was obliged to supply the absence of his wig, which was sent to the kitchen to be dried. After a time, the doctor exclaimed, with much animation, and with his accustomed lisp, "How very kind of you, my dear friend, to remember my love for *rothe goothe*." But his host, on going into the kitchen to ascertain the cause of so savoury a smell, found it was the doctor's wig smoking by the fire![19]

Here, as in many other cultural depictions of wigs from this period, the odd misfortunes arising from the seeming ill-suited nature of false hair for active modern life is a source of comic amusement. Despite complex nuances in terms of both lived reality and representation, there are overarching similarities regarding nineteenth-century attitudes to difference and concealment that make the study of these devices together important for the histories of disability, prostheses, and "passing," the divisive practice of concealing difference in order to appear normal, which I will turn to later.

Nineteenth-century discussions of artificial body parts often considered these technologies alongside one another. Commentaries on the expanding prosthesis trade in popular periodicals such as *Household Words*, *All the Year Round*, *Once a Week*, *Punch*, and *Tinsley's Magazine* discussed different types of prostheses comparatively.[20] In drawing our attention to the medical model that underpinned the nineteenth-century logic of prosthesis use, Jerrold concluded "Eyes Made to Order" as follows:

> It is a wise policy to remove from sight the calamities which horrify or sadden; and, as far as possible, to cultivate all that pleases from its beauty or grace. Therefore, let us shake our friend with the cork-leg by the hand, and,

> acknowledge that the imitation is worn in deference to our senses, receive it as a veritable flesh-and-blood limb; let us accept the wig of our unfortunate young companion, as the hair which he has lost; let us shut our eyes to the gold work that fastens the brilliantly white teeth of a young lady, whose natural dentition has been replaced; and, above all, let us never show, by sign or word, that the appearance of our friend (who has suffered tortures, and lost the sight of one eye) is changed after the treatment invented by M. Boissonneau.[21]

For Jerrold, all of the prostheses listed are linked in the way that they try to produce a "pleasing personnel."[22] Humorous items also often presented different kinds of artificial body parts as interchangeable. A sardonic article in *Punch* encouraged readers to "give a friend in need, personal and pecuniary, a Christmas-Box in the shape of a set of artificial teeth, or the 'Guinea Jaw' of our friend the Dentist, or a glass eye, or a gutta percha nose, or a wooden leg."[23] Later, an ironic etiquette miscellany in the *Sporting Times* provided readers with the following tongue-in-cheek advice: "If you know that a man has a glass eye, or a wooden leg, or a wig, ... always refer to the circumstance on every possible occasion."[24]

While the general definition of prosthesis remains fairly broad—the *Oxford English Dictionary* defines it as "[a]n artificial body part, such as a limb, a heart, or a breast implant"—I choose here to focus specifically on devices that attempt to replicate the physical form or mimic the close appearance of the body part for which they are substitutes. I do, however, like Katherine Ott, recognize that "the line between assistive and prosthetic technology is more like a hyphen." Ott challenges the distinction often drawn between prosthetic and assistive technologies, writing, "Since all useful technology is assistive, it is peculiar that we stipulate that some devices are assistive while others need no qualification."[25] I endorse this view, but choose to focus on devices that stand in visibly for missing body parts, rather than those that enhance or supplement diminished sensory capacities—for example, spectacles and/or hearing aids—since the literary depictions of such technologies interact more fully with the social attitudes to the conspicuously aesthetic construction of physical wholeness.

Structure

Prosthetic Body Parts is divided into seven chapters. Chapters 2 and 3 set up several major themes, which are then examined in relation to influential social factors in Chaps. 4, 5, and 6. Chapter 2 explores the construction of the concept of physical wholeness and the way in which fears of physical loss were perpetuated. The second chapter addresses nineteenth-century contexts, such as changing understandings of the human condition, new models of work, and changes in legislation. The chapter also analyses literary texts that stimulated anxiety regarding the neurological impact of body loss, including Frederick Marryat's *Jacob Faithful* (1834) and Silas Weir Mitchell's "The Case of George Dedlow" (1866). I end by investigating how the burgeoning prosthesis market reinforced preferences for physical normalcy in advertisements as a means to exploit it.

Chapter 3 demonstrates how the contexts outlined in the previous chapter impacted conceptualizations of agency and ability in prosthesis narratives. My discussion examines how a power play between person and prosthetic part was often imagined in literary and cultural depictions of such technologies. By exploring the extent to which artificial body parts were seen to enhance or assume the agency of the user, I argue that several prosthesis narratives produced transgressive prosthesis users or false body parts that threatened the dominance of the physically whole. Underlining the enduring nature of such themes, I analyse sources from across the historical scope of this project, including several sketches and short stories that appeared in publications such as *Kind Words*, *All the Year Round*, and *Longman's Magazine*; Poe's short story "The Man That Was Used Up"; Melville's novel *Moby-Dick*; Hood's narrative poem *Miss Kilmansegg and Her Precious Leg*; Frances Parker's illustrated narrative poem *The Flying Burgermaster* (1832); and Blackton's short film *The Thieving Hand*.[26]

In Chap. 4, I concentrate on the intersections between prosthesis use and social mobility, challenging predominant utopian views regarding nineteenth-century prosthetics. Centring on a case study of Dickens's popular portrayal of the villainous wooden-leg user Silas Wegg in *Our Mutual Friend*, I show how such works drew on anxieties surrounding the social position of amputees by presenting wooden-leg users as transgressive social climbers. I place Dickens's representation of Wegg in context with his other depictions of prosthesis users and those found in his journals *Household Words* and *All the Year Round*.[27] I also consider the cultural legacy of Wegg. This fourth chapter argues that stories such as *Our Mutual*

Friend problematized the logic of prosthesis use. Such tales suggested that, in an age of dominance of organic physical wholeness, prostheses were defunct when they failed to accurately mimic the appearance and function of the lost body part, and yet were ironically associated with fraudulence when successful.

By comparison, Chap. 5 traces representations of male and female prosthesis users in the marriage plot, the nineteenth-century narrative form most heavily populated by users of prosthetic devices. Building on the work of scholars such as Martha Stoddard Holmes and Talia Schaffer, this chapter identifies the prosthesis-marriage plot as a related yet separately identifiable formulaic narrative structure.[28] When viewed collectively, and at times also individually, prosthesis-marriage plots—including Hardy's novels *The Woodlanders* and *Jude the Obscure*, Dickens's *Barnaby Rudge* (1841) and *Dombey and Son*, and various short stories and sketches printed in publications including *Temple Bar*, *Fun*, *Cheshire Observer*, *Hearth and Home*, and *Chambers's Journal*—presented unstable affective and imaginative treatments of prosthesis users.[29] These representations shed light on the complex ways in which discourses of gender, class, and ableism intersected and how, in particular instances, the bodily status quo was brought into question or even outright rejected.

Chapter 6 investigates how ageing was a notable social factor scrutinized by prosthesis imaginaries. The cultural association of cosmetic prostheses (including wigs and false teeth) with ageing stems, at least in part, from satirical sources that paradoxically both bulwarked and mocked the hegemony of physical wholeness and youth. Stressing the extent to which preferences for youth were intertwined with demands for physical completeness, my analysis shows how the dominance of these two physical states was undermined by stories that either ridiculed the process of concealment for elderly users or presented unlikely ageing prostheticized heroes in unconventional ways. In this regard, I draw from genres that were in different ways invested in constructing bodily norms and deviances. I address the Gothic, by returning to Poe and his short stories "The Man That Was Used Up" and "The Spectacles" (1844), which are about sophisticated prostheses that onlookers find hard to detect. Thereafter, I turn to sensation fiction, by investigating Wilkie Collins's portrayals of wigs and dentures used by ageing characters in *Armadale*, *The Law and the Lady*, and *The Black Robe*. And then I focus on imperial adventure fiction, by analysing the unlikely past-their-prime prosthesis-using action heroes of Haggard's *King Solomon's Mines*, D. B. McKean's "A Wig and a

Wooden Leg" (1886), and the anonymously published "A Cure for Cannibalism" (1889).[30] Despite their differences in style and genre, collectively, these depictions of ageing prosthesis users challenged the dominance of physical wholeness and youth by laughing at the absurd results that demands for both effected.

The concluding seventh chapter turns to the British television network Channel 4's "Superhumans Return" (2015) advertising campaign for its coverage of the 2016 Paralympic Games as a case study.[31] By analysing video advertisements from this campaign, I highlight the way that contemporary sources interrogate a privileging of normalcy while remaining encoded by certain ableist inclinations. I then synthesize the various strands of the book's argument to make the case that the literary history of prosthesis is rich, complicated, and conflicted.

Prosthetic Body Parts builds on and adds nuance to historical work that traces the social construction of physical normalcy, a concept that I show was buttressed by an understanding of the healthy body as whole.[32] Like Lennard J. Davis, I explore the denigration of physical difference that such a rise encouraged. The prosthesis industry, which saw tremendous development in the nineteenth century, cashed in on the increasing mandate for physical normalcy. While contemporary journalism and advertising often lauded the accomplishments of an emerging group of professional prosthesis makers, many cultural and literary sources provided the other side of the picture, revealing the stereotypes, stigma, scepticism, inadequacies, and injustices attached to the use and dissemination of prosthetic devices. Victorian prosthesis narratives therefore complicated the hegemony of normalcy that Davis ascribes to this period. Nineteenth-century prosthesis narratives, though presented in a predominantly ableist and sometimes disablist manner, challenged the dominance of physical completeness as they questioned the logic of prostheticization or presented non-normative subjects in threateningly powerful ways.

Scope

To evidence the extent to which the nineteenth and early twentieth centuries were notable for their developments in real-life prosthetic technologies, I would like to briefly outline some key advances relevant to each of the devices investigated in the chapters that follow. Artificial limbs, in particular legs, saw significant transformation during the nineteenth century. Before 1830, the makers of artificial limbs—as in devices that attempted to replicate both the appearance and the function of a limb that had been lost—were few and far between. Rudimentary peg legs, tapered wooden posts upon which users could rest their amputated stumps, had been in use for centuries and remained the most popular replacements for lost limbs. James Potts made what is often considered the first modern artificial leg in 1816 when he supplied Henry William Paget, Lord Uxbridge, with a prosthetic replacement for the leg that he famously lost in the Battle of Waterloo.[33] Potts's prosthetic was hailed as a great success by Uxbridge, who was newly titled as the Marquess of Anglesey. The prosthesis came to be known interchangeably as both the Anglesey and the "clapper" leg—"so called because locomotion was accompanied by a clapping sound."[34] Paul Youngquist explains what made Potts's device special:

> Unlike the familiar peg leg, whose crude artificiality materialized the blunt claims of patriotism on the bodies of commoners, Anglesey's leg was lifelike and elegantly sculpted. It embodied a much more intimate fit between man and nation. And it allowed greater ease of mobility, communicating enough limp to mark the hero, while concealing enough stump to confirm the gentleman.[35]

Though certainly a major innovation, as Youngquist notes, the general circulation of the Anglesey leg was restricted by its high cost. The Anglesey design was replicated and made slightly more affordable on both sides of the Atlantic after Potts's death, first by two of his apprentices, Frederick Gray and William Selpho, and later by their imitators, competitors, and entrepreneurial protégés. It was not until the American Civil War, however, that such sophisticated prosthetic devices became more widely available. Guy Hasegawa's *Mending Broken Soldiers* (2012) documents the complex process that led to state provisions being provided to veterans for the purchase of artificial limbs. Before and especially after the American Civil War, many of the century's major innovations in lower-limb prosthesis took place on that side of the Atlantic. Benjamin Frank Palmer of Philadelphia won first prize at the International Exhibition of 1851 in London for his artificial leg, which used a spring in the foot to give firmness of step. In 1858 Douglas Bly developed what he called the "anatomical leg," which incorporated an ivory ball in a vulcanized rubber socket to provide polycentric ankle motion. Three years later, New Yorker A. A. Marks introduced the rubber foot, which simplified ankle joint manufacture and enabled a more lifelike gait. And in 1863, another New Yorker,

Dubois Parmelee, pioneered using atmospheric pressure as found in a suction socket to attach above-the-knee artificial legs.[36] The growth of the limb-prosthesis industry in this period owed much to developments in surgical practice, hygiene, and pain relief. Innovations such as the introduction of the Syme's method of amputation at the ankle joint, the introduction of anaesthetics such as ether and chloroform in the late 1840s, and the gradual adoption of Listerian principals of prophylactic antisepsis from the 1870s meant that more patients survived amputations and more survived with serviceable stumps suitable for being fitted with prosthetics as the century progressed.[37]

Developments in artificial arms were not nearly as impressive as the innovations in artificial legs. Sue Zemka explains that due to difficulties replicating the complex biomechanics of the human hand, artificial arms "languished on an impasse between functionality and a natural appearance."[38] Rudimentary hooks, available many years before the Victorian period, remained the most effective artificial hands up until and far beyond 1901—due to the limited availability of cybernetic artificial hands in our own time, one could even make the argument that devices of a very similar design remain the most effective replacements for missing hands today. Though, as Zemka states, one must be careful regarding the application of labels of "progress" and "improvement" to the nineteenth-century history of artificial arms, there certainly was growth. The improvements in artificial arms were insubstantial but the transatlantic expansion of the limb-prosthesis trade was unprecedented. For instance, in the 1820s, there were three artificial limb firms in London; by the 1880s, there were eighteen.[39]

While major innovations in artificial arms failed to materialize, ocular prostheses underwent major technological developments. In the 1840s, when the anatomy of the eye became more accurately understood, thanks to the work of ophthalmologists such as Amédée Bonnet, surgeons engineered a new, safer method of performing enucleation—the removal of the entire eyeball.[40] By cutting the four rectus muscles, which control eye movement, surgeons effected easier and more practical methods for extracting the globe. Later in the century, ophthalmic surgeons developed procedures for implanting support spheres that would give a better outcome to the placement of the artificial eye. The Mule's operation was the most popular of such procedures. The delivery of these operations was made more practical by the introduction of anaesthesia and prophylactic antisepsis. Artificial eyes themselves had been in use in modern Europe since the sixteenth century when pioneering French surgeon Ambroise Paré fabricated a covered and painted metal plate that could be worn over the eyelid of a lost eye. In the 1700s, the industry was dominated by Venice's talented glass blowers. But in 1822 France returned as the global centre for artificial eyes following the Boissonneau family's production of the first enamel artificial eye. Auguste Boissonneau's eyes dominated the European market in the 1840s, 1850s, and 1860s.[41]

Such market supremacy, however, was not long lived. German ocularishts started using cryolite glass in the 1870s. This material proved easy to work with and "finished to a more lifelike, opalescent shine," leading to the dominance of German (especially Wiesbaden) eyes.[42] Another important technological development spearheaded by German makers was the development of the "reform" or Snellen eye. Named after Dutch ophthalmologist Hermann Snellen, who called for artificial eyes suitable for enucleated sockets to be developed, the reform eye was created by the Müller-Uri family. As Ott writes: "Patients and ocularists preferred the Snellen design because it reduced the sunken appearance of the orbit and socket area of the face."[43]

Artificial teeth also saw major developments, especially in America. The introduction of anaesthesia in the 1840s meant that "[n]umerous people who had preferred tooth ache to the torture of extraction were now hastening to have rotten teeth cleared from their mouths."[44] Significant innovations followed, including the implementation of sulphur-hardened rubber—vulcanite—as a material for moulding bases. The use of this material significantly lowered the cost of false teeth, inaugurating what dental historian M. D. K. Bremner has called the era of "false teeth for the millions."[45] Earlier in the century, spring-less upper and lower sets began to appear. Though not necessarily a new idea (influential eighteenth-century French dentist Pierre Fauchard made three upper sets able to stay in place without springs during his career), in 1848 the US Patent Office granted a patent on false teeth held in place by atmospheric pressure to a Connecticut confectioner.[46] The first efficient porcelain crowns and bridges appeared in the final quarter of the nineteenth century following the inventions of the first satisfactory dental cement (an oxyphosphate of zinc) in 1869 and the foot-operated dentist's drill in 1871.[47]

Wigs, relatively simple devices, saw little change in terms of technical sophistication. The popularity of artificial hair, however, was a social phenomenon. As the fashion for wearing it trickled down the social ladder, Britain imported a huge amount of artificial hair from Europe. According to Alexander Rowland, in 1851 England imported 10,862 pounds of human hair (which was used to make wigs) from France alone.[48] The penchant for hair additions, including false fronts, chignons, and tresses, extended from the 1850s through to the 1890s, peaking in the 1860s. The demand for artificial hair altered ideas about what constituted physical wholeness. For women especially, the whole and normal body temporarily became one embellished with artificial hair. At the mid-century especially, to lack artificial hair was to be physically incomplete. And yet those whose use of artifice was too obvious were ironically lambasted in cultural and literary texts.

Not only did the nineteenth century generate technological and commercial progress in prosthetic technologies, but it also witnessed a concomitant upsurge in discussions about and representations of these devices in contemporary print and visual culture. If we individually search the terms "artificial leg," "glass eye," "wig," and "false teeth"—arguably the most commonly used and visually recognizable prostheses of the modern era—on the ProQuest resource *British Periodicals* (Collections I, II, and III), a similar graphic is produced by each search conducted: mentions of the term grew after 1830, increasing immensely in line with developments in the manufacture and circulation of that prosthesis towards the high-Victorian period, before reducing in number and eventually dropping off drastically after 1910.[49] In addition to the developments in prosthetic technologies, we can read the rise in discourse surrounding these devices through the Victorian period in relation to the upsurge in print culture, advertising, and marketing that was witnessed during this period.[50] Factors such as the reduction of newspaper stamp duty in 1836 and the abolition of advertisement duty in 1855 created a dramatic expansion in newspapers and magazines, providing greater space for fictional narratives including and advertisements for prostheses. The increase in interest surrounding prostheses in the 1830s also correlates with related historical factors, such as the 1834 Poor Law Amendment Act and Lambert Adolphe Quetelet's 1835 construction of "the average man," which had major consequences in terms of contemporary attitudes to disability.[51] Such events brought physical difference and the categorization of physical ability to the fore in an unprecedented way.

Critical Contexts

In arguing that Victorian prosthesis narratives challenged the hegemony of normalcy that was developing in the nineteenth century, I draw from important work in cultural and literary disability studies, Victorian studies, and literature and science.

The foundations for this project were laid by scholars such as Paul K. Longmore, Davis, Garland-Thomson, David T. Mitchell, Sharon L. Snyder, and Ato Quayson, who have demonstrated the importance of literature as cultural work that exposes and shapes attitudes to physical disability. One of the most influential and widely adapted frameworks to emerge from this field has been Mitchell and Snyder's theory of "narrative prosthesis."[52] By referring to the way that physical difference has been used throughout history as "a crutch upon which literary narratives lean for their representational power, disruptive potentiality, and analytical insight," Mitchell and Snyder build on the work of Longmore and others, arguing that "[d]isability inaugurates narrative, but narrative inevitably punishes its own prurient interests by overseeing the extermination of the object of fascination."[53] Quayson takes up a similar project in his book *Aesthetic Nervousness* (2007). He provides, however, a corrective to Mitchell and Snyder, arguing that disability often stimulates subliminal unease and moral panic, which is refracted within the structures of literature, a crisis he terms "aesthetic nervousness."[54] I provide a counterpoint to Mitchell and Snyder's argument about literature serving a prosthetic function in rendering physical difference invisible by showing that Victorian prosthesis narratives often brought physical difference to the fore, attacking the prosthetic part as an ineffective solution to functional and social issues related to physical difference and loss. For instance, in Grace Goldney's 1870 serialized novella *Marion's Choice* and William Henry Archibald Chasemore's 1878 *Judy* cartoon "Wicklebury's Wig," wigs are narratively and comically centred in order to ridicule their ill-suitedness.[55] By taking a more historicist approach than Mitchell and Snyder and Quayson, I consider how the rise of prosthetic technologies both effected and affected such depictions. As I also show, the complexity of disability representation is even thornier given the questions regarding the human-technology relationship that are evoked by the prosthetic body part.

The present study is also heavily indebted to the work of Vanessa Warne. Her essays "If You Should Ever Want an Arm" (2005), "Artificial Leg"

(2008), and "To Invest a Cripple with Peculiar Interest" (2009) reveal the significant position that prosthetic limbs held in the cultural imagination on both sides of the Atlantic in the nineteenth century.[56] Warne provides not only useful analyses of the discourses related to prosthesis representation—such as commercial enterprise, class privilege, and prosthetic compensation—but also guidance in terms of critical approach. As Warne observes: "The tension between literary representations and the lived experiences of amputees constitutes something of a problem for the study of prostheses in the Victorian period."[57] Related to Warne's concerns regarding the tension between fiction and reality, several scholars have been vocal in calling for a return to thinking about prosthetics literally rather than figuratively. Vivian Sobchack, a prominent media theorist and social critic, as well as a person with lived experience of limb-prosthesis use, writes:

> [T]he primary context in which "the prosthetic" functions literally rather than figuratively has been left behind—as has the experience and agency of those who, like myself, actually use prostheses without feeling "posthuman" and who, moreover, are often startled to read about all the hidden powers that their prostheses apparently exercise both in the world and in the imaginations of cultural theorists. Indeed, most of the scholars who embrace the prosthetic metaphor far too quickly mobilize their fascination with artificial and "posthuman" extensions of "the body" in the service of a rhetoric (and in some cases, a poetics) that is always located elsewhere—displacing and generalizing the prosthetic before exploring it first on its own quite extraordinary complex, literal (and logical) ground[.][58]

Similarly, Steven L. Kurzman argues, "[t]he major flaw with retroactively basing the prosthesis metaphor in artificial limbs is that it reinscribes the latter to support the model. It misrepresents artificial limbs as semi-autonomous agents, which I do not believe reflects the reality of how amputees relate to or use artificial limbs in either individual or social senses."[59] My analyses acknowledge Warne's concern about fiction versus reality while following Sobchack's and Kurzman's respective prompts to analyse the prosthesis *as prosthesis*.

A significant proportion of historical work on prosthesis focuses on military contexts and the provision of prostheses to veteran amputees.[60] Another notable trend linked to this work has been a focus on male users. For example, Katherine Ott, David Serlin, and Stephen Mihm's seminal collection of essays on the history of prostheses, *Artificial Parts, Practical Lives: Modern Histories of Prosthetics* (2002), investigates female prosthesis users in just two of its twelve chapters, while four of its essays are explicitly about war veterans. In her introduction to the volume, Ott herself acknowledges the limitations of the volume and notes that the anthology "is intended to stimulate research and critical inquiry into questions about … the gender dynamics of prostheses."[61] This task has been taken up recently by scholars including Luna Dolezal and Clare Stainthorp, who respectively investigate how recent media portrayals of the Paralympic athlete, actress, and model Aimee Mullins ultimately promote "possessive individualism … and the most banal patriarchal tendencies of mainstream consumerism," and how Victorian doctor and prosthetist Henry Robert Heather Bigg's 1885 book *Artificial Limbs and the Amputations Which Afford the Most Appropriate Stumps in Civil and Military Surgery* asserted a "professional and masculine agency to make the woman's body assume the position of something beheld rather than embodied."[62] My essay "Get the Best Article in the Market" also brings female prosthesis users into focus, revealing how particular literary texts used in advertisements and print media promoted the concealing ability of particular prosthetic devices to female users while warning them away from others.[63]

A major way that my work differs from much historical work on prosthesis is in terms of approach. To date a lot of historical research on prostheses has focused on the perceived successes of these devices without examining the normalizing forces that stimulated their development. Studies by scholars such as Erin O'Connor, Edward Steven Slavishak, and Guy Hasegawa usefully unpack the symbolic and functional value of artificial legs in nineteenth-century Britain and America—for instance, O'Connor identifies that "Prosthetics figured in the Victorian imagination as the closural movement of amputation, putting an end to the body's unsettling counter-narrative by materially effacing it as such"—but what needs to be probed further in relation to these devices and other forms of prosthesis is the problematic social mandate that, in part, brought about their proliferation.[64] I do not wish to imply that there were not benevolent agendas at heart in the development of prostheses in this period. Indeed, it is true that some prosthetists (including the British maker of devices for arm amputees George Webb Derenzy, the American artificial limb makers B. Frank Palmer and James A. Foster) had lived experience as amputees and developed prostheses, in part, to improve the lives of not only themselves but also others living with similar differences.[65] Rather, I believe that it is important to consider the ableism underpinning the demand for life-like prosthesis that could enable users—who were more often than not financially privileged—to pass as normal. In other words, I support Sarah Jain's observation that "the unspecified deficiency, the generalized defect or absence seems to naturalize the general form of the prosthesis and of the body alike. If the prosthesis presumes an enhancement to the 'natural' body in this account, then bodies and prostheses are already naturalized rather than being understood as socially constructed."[66] It is important for historians of prostheses to interrogate the naturalizing of physical loss as deficiency. In adopting a social-constructivist view of prostheses, I do not wish to deny or overlook the physical difficulties, pain, and mental anguish occasioned by losing or not being born with a particular body part, but I do wish to show that such issues have been exacerbated by social conditions that have valourized physical wholeness and denigrated bodies deemed incomplete. It is idealistic to think that prostheses were produced solely to make the lives of physically different people better. While this is no doubt an important part of the equation, it should be acknowledged that the very existence of these devices was predicated by a privileging of normative looks, functions, and movement patterns.

Related to the history of prosthesis, *Prosthetic Body Parts* also contributes to an emerging historiography of passing, a practice that in the context of disability "refers to the way people conceal social markers of impairment to avoid the stigma of disability and pass as 'normal.'"[67] Despite the clear links with prosthesis use, a kind of supplementing of the body underpinned by a medical approach invested in materially effacing the supposedly "fixable" issue of bodily loss, surprisingly little historical work on prosthesis directly addresses the practice of passing.[68] Jeffrey A. Brune and Daniel J. Wilson explain how passing is a contested practice in disability studies since it "can take a psychological toll [on those who attempt to 'pass'] and can also reinforce—or, at least, fail to challenge—the stigma of disability." They also, however, note: "Even when passing seems to reinforce the stigma of disability, it is more productive, and more just, to challenge the ableism that compels people to pass rather than blame the individuals who choose to do so."[69] By exploring attitudes to passing through literature and culture, I explore conflicting social attitudes to this mode of self-presentation, moving beyond the current (yet also important) penchant for investigating the personal perspective of the passing subject.

Moreover, *Prosthetic Body Parts* intervenes in debates surrounding the human-technology relationship in the nineteenth and early twentieth centuries, a topic that has seen considerable interest within the burgeoning field of literature and science—not the least because, as Laura Otis observes, "[t]hrough their comparisons of bodies and machines, [nineteenth-century] scientists and literary writers contributed to a new cultural understanding of selfhood."[70] Unlike previous scholarship, I show that within the literary imagination the complex dynamic of the human-prosthesis relationship challenged not only subject/object binaries but also the cultural dominance of organic physical wholeness. Tamara Ketabgian, who in *The Lives of Machines* uses the metaphor of prosthesis to discuss the complex subject-object relations between man and industrial machine, argues, "Victorian machines were not simply soulless, lifeless, predictable, and unidimensional; not simply opposed to organic feeling and vitality; and not simply reductive material objects—if objects are ever so."[71] Elsewhere, Katharina Boehm's edited collection *Bodies and Things in Nineteenth-Century Literature and Culture*, which draws heavily on Bill Brown's work on thing theory, presents the argument that the subject and the object were not always oppositional in the nineteenth century; instead they connected through "networked and processual relationships."[72] Adding to this work on the interfaces and ontological overlaps of the human and the machine, the subject and the object, this book draws needed attention to the prosthetic body part, a device that (perhaps more than any other) raises questions about where the subject ends and the object begins.

Language

Because of the extent to which acceptable language is a contested topic in disability studies—particularly when dealing with historical sources that use terms that we now consider offensive and/or derogatory—in writing about prosthesis users I have had to make careful decisions about terminology. I primarily use the term *disabled* when discussing those perceived to be *missing* limbs. Often, I use the more specific term *amputee*. It is true that the term *disabled* was used infrequently to describe people with physical impairments prior to the First World War, but this term is more neutral than the alternatives used in Victorian times.[73] I avoid using terms such as *afflicted*, *defective*, *infirm*, and *cripple* unless writing from the perspective of individuals from the nineteenth or early twentieth century. When I do use these terms, they appear in quotation marks to show that they are not my own. However, because this study does not deal with disability alone—it would be misleading to call those deemed to be *missing* hair, teeth, or even an eye disabled despite the stigma, and, at times, the functional difficulties accompanying their physical conditions—I tend to use provocative terms such as *incomplete* and *disaggregated* to describe those considered to be *lacking* body parts. Though these words were not commonly used in the period under discussion, and they certainly do not express any personal bias as regards an idealized or *normative* vision about how the body should appear, they encapsulate the hegemonic and problematic (though not exclusive) attitude to perceived physical *losses* often exhibited in nineteenth- and early twentieth-century sources.

These terms are certainly unsettling, and it is important to note that the attitude that they express regarding bodily difference is a socially constructed one, based on the notion that during the nineteenth century the *normal* or *physically complete/whole* body was the dominant paradigm. When describing bodies that would have been considered *non-normative*, I use terms such as *loss* and *missing* though I would like to acknowledge here that I am uncomfortable with the homogenizing view of difference as *lack*. The term *whole*, which was often used during the nineteenth century when describing the *normal* body (also a social construction), is used alongside its synonym *complete*—a term less commonly deployed in such context in the nineteenth century—for linguistic variety. *Physical integrity* is another variation that I employ to avoid repetition. In identifying the dominant social position of those who were deemed to exhibit *wholeness*, I also occasionally borrow Garland-Thomson's provocative term *normate*. As Garland-Thomson herself explains:

> This neologism names the veiled subject position of cultural self, the figure outlined by the array of deviant others whose marked bodies shore up the normate's boundaries. The term normate usefully designates the social figure through which people can represent themselves as definitive human beings. Normate … is the constructed identity of those who, by way of the bodily configurations and cultural capital they assume, can step into a position of authority and wield the power it grants them. If one attempts to define the normate position by peeling away all the marked traits within the social order at this historical moment, what emerges is a very narrowly defined profile that describes only a minority of actual people.[74]

The concept *normate* therefore aptly encapsulates the hegemonic yet constructed identity held by those believed to display *wholeness*—in reality a minority, whose very state of *completeness* was ever subject to change. The fact is that even today, over one hundred years of medical progress later, relatively few of us remain *normatively* and *organically whole* over an entire life course—though we may think of and perceive our bodies as *whole* regardless of how they are received by others. The overwhelming majority of us *lose*, or are not born with, at least one body part, however minor it might seem to us. In the nineteenth century, hair, teeth, limbs, and eyes were among the body parts most at risk. The key task for *Prosthetic Body Parts* is to show how our literary and cultural history reveals that attempts to conceal physical differences have not always been privileged.

Notes

1. Herman Melville, *Moby-Dick* (London: Penguin, 2012); Edgar Allan Poe, "The Man That Was Used Up: A Tale of the Late Bugaboo and Kickapoo Campaign," in *The Works of the Late Edgar Allan Poe*, vol. 4, by Edgar Allan Poe (New York: Blakeman & Mason, 1859); Charles Dickens, *Dombey and Son* (London: Penguin, 2002); and Charles Dickens, *Our Mutual Friend* (London: Everyman, 2000). Long John Silver (from Robert Louis Stevenson's *Treasure Island* [1882]) and Captain Hook (from J. M. Barrie's *Peter Pan* [1904]) may seem like conspicuous omissions, but I address these in the following essay: Ryan Sweet, "Pirates and Prosthetics: Manly Messages for Managing Limb Loss in Victorian and Edwardian Adventure Narratives," in *The Victorian Male Body: The Diverse Embodiment of White Masculinity in the Nineteenth Century*, eds. Joanne Parsons and Ruth Heholt (Edinburgh: Edinburgh University Press, 2018). As I note in this piece, contrary to popular belief, Stevenson in fact makes it clear that Silver is a crutch, rather than an artificial limb, user.
2. Wilkie Collins, *Armadale* (Oxford: Oxford University Press, 1999); Wilkie Collins, *The Law and the Lady* (Oxford: Oxford University Press, 2008); Wilkie Collins, *The Black Robe* (London: Chatto and Windus, 1910); Thomas Hardy, *The Woodlanders* (Oxford: Oxford University Press, 2009); Thomas Hardy, *Jude the Obscure* (Oxford: Oxford University Press, 2008); H. Rider Haggard, *King Solomon's Mines*, Project Gutenberg, last modified October 12, 2012. https://www.gutenberg.org/files/2166/2166-h/2166-h.htm; and Arthur Conan Doyle, *The Sign of Four* (Garches: Feedbooks, n.d).

3. J. Stuart Blackton, *The Thieving Hand*, Vitagraph Company of America, 1908, YouTube, accessed June 16, 2018, https://www.youtube.com/watch?v=J2Dr0MBWXgQ.
4. Erin O'Connor, *Raw Material: Producing Pathology in Victorian Culture* (Durham: Duke University Press, 2000); Jennifer Esmail, *Reading Victorian Deafness: Signs and Sounds in Victorian Literature and Culture* (Athens: Ohio University Press, 2013); Claire L. Jones, ed. *Rethinking Modern Prostheses in Anglo-American Commodity Cultures, 1820–1939* (Manchester: Manchester University Press, 2017).
5. Guy R. Hasegawa, *Mending Broken Soldiers: The Union and Confederate Programs to Supply Artificial Limbs* (Carbondale: Southern Illinois University Press, 2012), 50–51.
6. Gordon Phillips, *Best Foot Forward: Chas. A. Blatchford & Sons Ltd. (Artificial Limb Specialists) 1890–1990* (Cambridge: Granta, 1990), 30.
7. Fernando Galván, "Poe versus Dickens: An Ambiguous Relationship," in *A Descent into Edgar Allan Poe and His Works: The Bicentennial*, eds. Beatriz González Moreno and Margarita Rigal Aragón (Oxford: Peter Lang, 2010), 3–4.
8. Ryan Sweet, "'Get the best article in the market': Prostheses for Women in Nineteenth-Century Literature and Commerce," in *Rethinking Modern Histories of Prostheses in Anglo-American Commodity Cultures, 1850–1960*, ed. Claire L. Jones (Manchester: Manchester University Press, 2017).
9. André de Blaumont, "My Fiancé's Glass Eye," in *Short Stories: A Magazine of Select Fiction*, ed. Alfred Ludlow White, trans. Mrs. Huntington Denton, vol. XV (New York: The Current Literature Publishing Co., 1894).
10. Andrew Wynter, "The Artificial Man," *Once a Week* 1, no. 11 (1859): 220.
11. John Brown, "A Memoir of Robert Blincoe, an Orphan Boy," in *Factory Lives: Four Nineteenth-Century Working-Class Autobiographies*, ed. James R. Simmons Jr. (London: Broadview, 2007); William Dodd, *A Narrative of the Experience and Sufferings of William Dodd, A Factory Cripple. Written by Himself*, in *Factory Lives: Four Nineteenth-Century Working-Class Autobiographies*, edited by James R. Simmons, Jr. (London: Broadview, 2007); Henry Mayhew, *London Labour and the London Poor: A Cyclopaedia of the Condition and Earnings of Those That Will Work, Those That Cannot Work, and Those That Will Not Work*, vol. 1 (London: Griffin, Bohn, and Company, 1861).
12. I use the modifier "primarily" here as ocular prostheses and dentures, though used mostly for aesthetic reasons, do also serve functional purposes. As popular mid-century ocularist Auguste Boissonneau explained, "When the atrophied globe no longer exactly fills the socket, the eyelids fall one over the other, and the irritating contact of the lashes with the palpebral mucous membrane gives rise to inflammation[.] … The adapta-

tion of an artificial eye, by sustaining and separating the eyelids, and keeping them in their natural position, will prevent or cure such painful affections." Auguste Boissonneau, *General Observations on Artificial Eyes, Their Adaption, Employment and the Means of Procuring Them* (Paris: J.-B. Baillière and Son, 1862), 7–8. Of the functional purposes of false teeth, "though adequate mastication could be achieved" with the hardened gums left after teeth and projecting stumps had departed, "over closing of the lower jaw sometimes brought on deafness." A well-fitting set of false teeth could remove this risk. John Woodforde, *The Strange Story of False Teeth* (London: Routledge and Kegan Paul, 1968), 87.

13. "Women. By Our Office Boy," *Pick-Me-Up*, no. 194 (1892): 189.
14. Pamela Horn notes that vanity was associated with moral corruption and could lead to one's fall down the social ladder: "Middle-class moralists feared that girls preoccupied with fashion might slip into prostitution in order to obtain the cash needed to appear stylish. William Acton, in discussing the cause of prostitution in the late 1850s, maintained that 'vanity, vanity and then vanity' was the prime factor." Pamela Horn, *Pleasures and Pastimes in Victorian Britain* (Stroud: Amberley Publishing, 1999), 45.
15. William Blanchard Jerrold, "Eyes Made to Order," *Household Words* 4, no. 81 (1851): 64.
16. "Mr. Wegg and His Class," *All the Year Round* 18, no. 441 (1877).
17. "Mr. Aked Denounces Wigs," *The Penny Illustrated Paper*, July 27, 1907, 56.
18. Hubert von Herkomer, *The Last Muster*, 1875, oil on canvas, 214.5 × 159 cm, Lady Lever Art Gallery, Liverpool.
19. William Carew Hazlitt, *The New London Jest Book* (London: Reeves and Turner, 1871), 28.
20. See, for example, "The Eyes of the World," *Punch* 3 (1842); Jerrold, "Eyes Made to Order"; A. W., "The Artificial Man," *Once a Week* 1, no. 11 (1859); "Bodily Repairs," *Once a Week* 2, no. 38 (1868); and "Wigs," *All the Year Round* 9, no. 225 (1873).
21. Jerrold, "Eyes Made to Order," 66.
22. Ibid., 66.
23. "Christmas Boxes for Beauty," *Punch*, January 13, 1872, 19.
24. "Etiquette," *Sporting Times*, June 9, 1883, 1.
25. Katherine Ott, "The Sum of Its Parts: An Introduction to Modern Prosthetics," in *Artificial Parts, Practical Lives: Modern Histories of Prosthetics*, eds. Katherine Ott, David Serlin, and Stephen Mihm (New York: New York University Press, 2002), 21.
26. "A Romance of War," *Kind Words* 1 May 1872; "Peg-Legged Bob," *All the Year Round* 11, no. 280 (1874); M. E. Francis, "A Rustic Argus." *Longman's Magazine* 35, no. 210 (1900); Poe, "The Man That Was Used

Up"; Melville, *Moby-Dick*; Thomas Hood, *Miss Kilmansegg and her Precious Leg*, in *Selected Poems of Thomas Hood*, ed. John Clubbe (Cambridge: Harvard University Press, 1970); [Frances Parker], the Countess of Morley, *The Flying Burgermaster: A Legend of the Black Forest* (n.p.: F. Morley, 1832); Blackton, *The Thieving Hand*.

27. For example, [Richard H. Horne], "Dust; or Ugliness Redeemed," *Household Words* 1, no. 16 (1850); [Henry Morley], "Ground in the Mill," *Household Words* 9, no. 213 (1854); "Legs: Wooden and Otherwise," *All the Year Round* 14, no. 350 (1875); "Mr. Wegg and His Class."
28. Martha Stoddard Holmes, *Fictions of Affliction: Physical Disability in Victorian Culture* (Ann Arbor: University of Michigan Press, 2004); Talia Schaffer, *Romance's Rival: Familiar Marriage in Victorian Fiction* (Oxford: Oxford University Press, 2016).
29. Hardy, *The Woodlanders*; Hardy, *Jude the Obscure*; Charles Dickens, *Barnaby Rudge* (Oxford: Oxford University Press, 2003); Dickens, *Dombey and Son*; Buchanan, "Lady Letitia's Lilliput Hand"; "The Stricken Fawn," *Fun* 58, no. 1489 (1893); "He Fixed Her with His Glassy Eye," *Cheshire Observer*, April 5, 1879; G. W. C., "Good Advice and a Wooden Leg," *Hearth and Home: An Illustrated Weekly Journal for Gentlewomen*, no. 116 (1893); "The False Hair," *Chambers's Journal of Popular Literature, Science and Arts* 8, no. 396 (1861).
30. Poe, "The Man That Was Used Up"; Edgar Allan Poe, "The Spectacles," in *Prose Tales (Second Series) Arthur Gordon Pym*, by Edgar Allan Poe (Boston: Dana Estes, 1884); Collins, *Armadale*; Collins, *The Law and the Lady*; Collins, *The Black Robe*; Haggard, *King*; D. B. McKean, "A Wig and a Wooden Leg," *Chatterbox*, no. 4 (1886); "A Cure for Cannibalism," *Pick-Me-Up*, no. 37 (1889).
31. Channel 4, advertisement for Channel 4's coverage of the Paralympic Games 2016, YouTube, last modified, November 6, 2015, https://www.youtube.com/watch?v=1FKcmEXPnUw.
32. Lennard J. Davis, *Enforcing Normalcy: Disability, Deafness, and the Body* (London: Verso, 1995).
33. Potts patented an early version of this design in 1805. Phillips *Best Foot Forward*, 30.
34. Henry Paget, *One Leg: The Life and Letters of Henry William Paget, First Marquess of Anglesey, K.G. 1768–1854* (London: Leo Cooper, 1996), 153.
35. Paul Youngquist, *Monstrosities: Bodies and British Romanticism* (Minneapolis: Minnesota University Press, 2003), 184.
36. For more on the technological developments of these devices in the nineteenth century, see ibid., 32–8; John Kirkup, *A History of Amputations* (London: Springer-Verlag, 2007), 159–63; Hasegawa, *Mending*, 8–20.

37. The histories of such developments are by no means straightforward, however. Much controversy surrounded each of the innovations mentioned, hampering their widespread adoption. See, for instance, David Hamilton, "The Nineteenth-Century Surgical Revolution—Antisepsis or Better Nutrition?" *Bulletin of the History of Medicine* 56, no. 1 (1982); Martin S. Pernick, *A Calculus of Suffering: Pain, Professionalism and Anesthesia in Nineteenth Century America* (New York: Columbia University Press, 1985); Nicholas J. Fox, "Scientific Theory Choice and Social Structure: The Case of Joseph Lister's Antisepsis, Humoral Theory and Asepsis," *History of Science* 26, no. 4 (1988); Christopher Lawrence and Richard Dixey, "Practising on Principle: Joseph Lister and the Germ Theories of Disease," in *Medical Theory, Surgical Practice: Studies in the History of Surgery*, ed. Christopher Lawrence (London: Routledge, 1992); Lindsay Granshaw, "'Upon This Principle I Have Based a Practice': The Development and Reception of Antisepsis in Britain, 1807–1890," in *Medical Innovations in Historical Perspective*, ed. John V. Pickstone (Basingstoke: Palgrave, 1992); Kirkup, *A History*; and Jacqueline H. Wolf, *Deliver Me from Pain: Anesthesia and Birth in America* (Baltimore: Johns Hopkins University Press, 2011).
38. Sue Zemka, "1822, 1845, 1869, 1893, and 1917: Artificial Hands," *BRANCH: Britain, Representation and Nineteenth-Century History*, ed. Dino Franco Felluga [2015], 2, http://www.branchcollective.org/?ps_articles=sue-zemka-1822-1845-1869-1893-and-1917-artificial-hands.
39. Phillips, *Best*, 34–35.
40. Bonnet was an influential orthopaedic surgeon, who was a prominent figure at Lyon's Hôtel-Dieu Hospital. Crucially, Bonnet re-described the Tenon's capsule in 1841, which though earlier identified by Jacques Tenon, proved an elusive aspect of the eye's anatomy up until this point. Katherine Ott, "Hard Wear and Soft Tissue: Craft and Commerce in Artificial Eyes," in *Artificial Parts, Practical Lives: Modern Histories of Prosthetics*, eds. Katherine Ott, David Serlin, and Stephen Mihm (New York: New York University Press, 2002), 166n9.
41. Ibid., 151–52.
42. Ibid., 152–55.
43. Ibid., 154.
44. Woodforde, *Strange Story of False Teeth*, 88.
45. Maurice David Kaufman Bremner, *The Story of Dentistry from the Dawn of Civilization to the Present* (New York: Dental Items of Interest, 1939), 107–11.
46. Woodforde, *Strange Story of False Teeth*, 72.
47. Ibid., 76–77.

48. Alexander Rowland, *The Human Hair, Popularly and Physiologically Considered with Special Reference to Its Preservation, Improvement and Adornment, and the Various Modes of Its Decoration in All Countries* (London: Piper, Brothers & Co., 1853), 164.
49. ProQuest, *British Periodicals*, accessed July 5, 2018, https://www.proquest.com/products-services/british_periodicals.html.
50. For more on the growth of advertising, see Thomas Richards, *The Commodity Culture of Victorian England: Advertising and Spectacle, 1851–1914* (Stanford: Stanford University Press, 1990); Terence Nevett, "Advertising," in *Victorian Periodicals and Victorian Society*, eds. J. Don Vann and Rosemary T. Van Arsdel (Toronto: University of Toronto, 1994); and Roy Church, "Advertising Consumer Goods in Nineteenth-Century Britain: Reinterpretations," *The Economic History Review* 53, no. 4 (2000). For more on the development of Victorian print culture, see Richard Daniel Altick, *The English Common Reader: A Social History of the Mass Reading Public, 1800–1900* (Columbus: Ohio State University Press, 1998); John Sutherland, *Victorian Novelists and Publishers* (London: Athlone Press, 1976); Simon Eliot, *Some Patterns and Trends in British Publishing 1800–1919* (London: Bibliographical Society, 1994). For more on the marketing of medical devices, see Bruce Haley, *The Healthy Body and Victorian Culture* (Cambridge: Harvard University Press, 1978); Anne Digby, *Making a Medical Living: Doctors and Patients in the English Market for Medicine, 1720–1911* (Cambridge: Cambridge University Press, 2002); James F. Stark, "'Recharge My Exhausted Batteries': Overbeck's Rejuvenator, Patenting, and Public Medical Consumers, 1924–1937," *Medical History* 58, vol. 4 (2014); Claire L. Jones, *The Medical Trade Catalogue, 1870–1914* (London: Pickering & Chatto, 2014); Claire L. Jones, ed. *Rethinking Modern Prostheses in Anglo-American Commodity Cultures, 1820–1939* (Manchester: Manchester University Press, 2017).
51. Davis, *Enforcing Normalcy*; Holmes, *Fictions of Affliction.*
52. David T. Mitchell and Sharon L. Snyder, *Narrative Prosthesis: Disability and the Dependencies of Discourse* (Ann Arbor: University of Michigan Press, 2000).
53. Ibid., 57.
54. Ato Quayson, *Aesthetic Nervousness: Disability and the Crisis of Representation* (New York: Columbia University Press, 2007).
55. Grace Goldney, *Marion's Choice*, *Aunt Judy's Magazine* 8, no. 112–14 (1870); William Henry Archibald Chasemore, "Wicklebury's Wig," *Judy*, January 2, 1878.
56. Vanessa Warne, "'If You Should Ever Want an Arm': Disability and Dependency in Edgar Allan Poe's "The Man That Was Used Up,"' *Atenea* 25, no. 1 (2005); Vanessa Warne, "Artificial Leg," *Victorian Review* 34,

no. 1 (2008); Vanessa Warne, "'To Invest a Cripple with Peculiar Interest': Artificial Legs and Upper-Class Amputees at Mid-Century," *Victorian Review* 35, no. 2 (2009).
57. Warne, "Artificial Leg," 32.
58. Vivian Sobchack, "A Leg to Stand On: Prosthetics, Metaphor, and Materiality," in *The Prosthetic Impulse: From a Posthuman Present to a Biocultural Future*, eds. Marquard Smith and Joanne Morra (London: MIT Press, 2006), 20.
59. Steven L. Kurzman, "Presence and Prosthesis: A Response to Nelson and Wright," *Cultural Anthropology* 16, no. 3 (2001): 378. Also see Sarah S. Jain, "The Prosthetic Imagination: Enabling and Disabling the Prosthesis Trope," *Science, Technology, & Human Values* 24, no. 1 (1999): 31–54; and Ott, "The Sum," 1–3.
60. See, for example, Phillips, *Best*; Lisa Herschbach, "Prosthetic Reconstruction: Making the Industry, Re-Making the Body, Modelling the Nation," *History Workshop Journal* 44 (1997); David D. Yuan, "Disfigurement and Reconstruction in Oliver Wendell Holmes's 'The Human Wheel, Its Spokes and Felloes,'" in *The Body and Physical Difference: Discourses of Disability*, eds. David T. Mitchell and Sharon L. Snyder (Ann Arbor: University of Michigan Press, 1997; Erin O'Connor, "'Fractions of Men': Engendering Amputation in Victorian Culture," *Comparative Studies in Society and History* 39, no. 4 (1997); Connor, *Raw Material*; R. B. Rosenburg, "'Empty Sleeves and Wooden Pegs': Disabled Confederate Veterans in Image and Reality," in *Disabled Veterans in History*, ed. David A. Gerber (Ann Arbor: University of Michigan Press, 2012); Youngquist, *Monstrosities*, 161–90; Katherine Ott, "Carnage Remembered: Prosthetics in the US Military since the 1860s," in *Materializing the Military*, eds. Bernard Finn and Barton C. Hacker (London: Cromwell, 2005); Neil Handley, "Artificial Eyes and the Artificialization of the Human Face," in *Devices and Designs: Medical Technologies in Historical Perspective*, eds. Carsten Timmermann and Julie Anderson (Basingstoke: Palgrave, 2006); Kirkup, *A History*, 155–72; and Hasegawa, *Mending*.
61. Ott, "The Sum," 7.
62. Luna Dolezal, "Representing Posthuman Embodiment: Considering Disability and the Case of Aimee Mullins," *Women's Studies* 46, no. 1 (2017): 74; Clare Stainthorp, "Activity and Passivity: Class and Gender in the Case of the Artificial Hand," *Victorian Literature and Culture* 45, no. 1 (2017): 2.
63. Sweet, "Get."

64. O'Connor, *Raw Material*, 105; Edward Steven Slavishak, "Artificial Limbs and Industrial Workers' Bodies in Turn-of-the-Century Pittsburgh," *Journal of Social History* 37, no. 2 (2003); and Hasegawa, *Mending*.
65. Laurel Daen, "'A hand for the one-handed': Prosthesis User-Inventors and the Market for Assistive Technologies in Early Nineteenth-Century Britain," in *Rethinking Modern Prostheses in Anglo-American Commodity Cultures, 1820–1939*, ed. Claire L. Jones (Manchester: Manchester University Press, 2017), 93; Caroline Lieffers, "Itinerant Manipulators and Public Benefactors: Artificial Limb Patents, Medical Professionalism and the Moral Economy in Antebellum America," in *Rethinking Modern Prostheses in Anglo-American Commodity Cultures, 1820–1939*, ed. Claire L. Jones (Manchester: Manchester University Press, 2017), 137; Hasegawa, *Mending*, 11.
66. Jain, "The Prosthetic," 39.
67. Jeffrey A. Brune and Daniel J. Wilson, eds., *Disability and Passing: Blurring the Lines of Identity* (Philadelphia: Temple University Press, 2013), 1.
68. The essays in Jones's *Rethinking Modern Prostheses* are notable exceptions.
69. Brune and Wilson, *Disability and Passing*, 4–5.
70. Laura Otis, *Literature and Science in the Nineteenth Century: An Anthology* (Oxford: Oxford University Press, 2002), 15.
71. Tamara Ketabgian, *The Lives of Machines: The Industrial Imaginary in Victorian Literature and Culture* (Ann Arbor: University of Michigan Press, 2011), 2.
72. Bill Brown, "Thing Theory," *Critical Inquiry* 28, no. 1 (2001); Katharina, Boehm, ed., *Bodies and Things in Nineteenth-Century Literature and Culture* (Basingstoke: Palgrave, 2012), 2.
73. For more on the etymology of this term and the politics of employing it in historical studies, see Nadja Durbach, *Spectacle of Deformity: Freak Shows and Modern British Culture* (Berkeley: University of California Press, 2009), 16–19.
74. Rosemarie Garland-Thomson, *Extraordinary Bodies: Figuring Physical Disability in American Culture and Literature* (New York: Columbia University Press, 1997), 8.

References

Altick, Richard Daniel. *The English Common Reader: A Social History of the Mass Reading Public, 1800–1900*. Columbus: Ohio State University Press, 1998.

Anderson, Julie. *War, Disability and Rehabilitation in Britain: Soul of a Nation*. Manchester: Manchester University Press, 2011.

A. W. "The Artificial Man". *Once a Week* 1, no. 11 (1859): 218–20.

Blackton, J. Stuart. *The Thieving Hand*. Vitagraph Company of America, 1908. YouTube. Accessed June 16, 2018. https://www.youtube.com/watch?v=J2Dr0MBWXgQ.
Blaumont, André de. "My Fiancé's Glass Eye." In *Short Stories: A Magazine of Select Fiction*. Edited by Alfred Ludlow White. Translated by Mrs. Huntington Denton. Vol. XV: 277–81. New York: The Current Literature Publishing Co., 1894.
"Bodily Repairs". *Once a Week* 2, no. 38 (1868): 226–29.
Boehm, Katharina, ed. *Bodies and Things in Nineteenth-Century Literature and Culture*. Basingstoke: Palgrave, 2012.
Boissonneau, Auguste. *General Observations on Artificial Eyes, Their Adaption, Employment and the Means of Procuring Them*. Paris: J.-B. Baillière and Son, 1862.
Bremner, Maurice David Kaufman. *The Story of Dentistry from the Dawn of Civilization to the Present*. New York: Dental Items of Interest, 1939.
Brown, Bill. "Thing Theory." *Critical Inquiry* 28, no. 1 (2001): 1–22.
Brown, Elspeth. "The Prosthetics of Management: Motion Study, Photography, and the Industrialized Body in World War I America." In *Artificial Parts, Practical Lives: Modern Histories of Prosthetics*, edited by Katherine Ott, David Serlin, and Stephen Mihm, 249–81. New York: New York University Press, 2002.
Brown, John. "A Memoir of Robert Blincoe, an Orphan Boy." In *Factory Lives: Four Nineteenth-Century Working-Class Autobiographies*, edited by James R. Simmons Jr, 87–180. London: Broadview, 2007.
Brune, Jeffrey A., and Daniel J. Wilson, eds. *Disability and Passing: Blurring the Lines of Identity*. Philadelphia: Temple University Press, 2013.
Buchanan, Robert Williams. "Lady Letitia's Lilliput Hand." *Temple Bar* 4 and 5 (1862): 551–69; 114–31.
Carey, John. *The Violent Effigy: A Study of Dickens' Imagination*. London: Faber and Faber, 1973.
Channel 4. Advertisement for Channel 4's coverage of the Paralympic Games 2016. YouTube. Last modified November 6, 2015. https://www.youtube.com/watch?v=1FKcmEXPnUw.
Chasemore, William Henry Archibald. "Wicklebury's Wig." *Judy*, January 2, 1878.
"Christmas Boxes for Beauty". *Punch*, January 13, 1872.
Church, Roy. "Advertising Consumer Goods in Nineteenth-Century Britain: Reinterpretations." *The Economic History Review* 53, no. 4 (2000): 621–45.
Collins, Wilkie. *Armadale*. Oxford: Oxford University Press, 1999.
———. *The Law and the Lady*. Oxford: Oxford University Press, 2008.
———. *The Black Robe*. London: Chatto and Windus, 1910.
"A Cure for Cannibalism". *Pick-Me-Up*, no. 37 (1889): 171.
"Curt Comments". *Fun*, no. 835 (1881): 187.

Daen, Laurel. "'A hand for the one-handed': Prosthesis User-Inventors and the Market for Assistive Technologies in Early Nineteenth-Century Britain." In *Rethinking Modern Prostheses in Anglo-American Commodity Cultures, 1820–1939*, edited by Claire L. Jones, 93–113. Manchester: Manchester University Press, 2017.

Davis, Lennard J. *Enforcing Normalcy: Disability, Deafness, and the Body*. London: Verso, 1995.

Dickens, Charles. *Barnaby Rudge*. Oxford: Oxford University Press, 2003.

———. *Dombey and Son*. London: Penguin, 2002.

———. *Our Mutual Friend*. London: Everyman, 2000.

Digby, Anne. *Making a Medical Living: Doctors and Patients in the English Market for Medicine, 1720–1911*. Cambridge: Cambridge University Press, 2002.

Dodd, William. *A Narrative of the Experience and Sufferings of William Dodd, A Factory Cripple. Written by Himself*. In *Factory Lives: Four Nineteenth-Century Working-Class Autobiographies*, edited by James R. Simmons, Jr., 181–300. London: Broadview, 2007.

Dolezal, Luna. "Representing Posthuman Embodiment: Considering Disability and the Case of Aimee Mullins." *Women's Studies* 46, no. 1 (2017): 60–75.

Doyle, Arthur Conan. *The Sign of Four*. Garches: Feedbooks, n.d.

Du Maurier, George. *The Martian*. North Hollywood: Aegypan, 2009.

Durbach, Nadja. *Spectacle of Deformity: Freak Shows and Modern British Culture*. Berkeley: University of California Press, 2009.

Eliot, Simon. *Some Patterns and Trends in British Publishing 1800–1919*. London: Bibliographical Society, 1994.

Esmail, Jennifer. *Reading Victorian Deafness: Signs and Sounds in Victorian Literature and Culture*. Athens: Ohio University Press, 2013.

"Etiquette". *Sporting Times*, June 9, 1883.

"The Eyes of the World". *Punch* 3 (1842): 68.

"The False Hair". *Chambers's Journal of Popular Literature, Science and Arts* 8, no. 396 (1861): [65]–67.

Fox, Nicholas J. "Scientific Theory Choice and Social Structure: The Case of Joseph Lister's Antisepsis, Humoral Theory and Asepsis." *History of Science* 26, no. 4 (1988): 367–97.

Francis, M. E. "A Rustic Argus." *Longman's Magazine* 35, no. 210 (1900): 521–34.

Galván, Fernando. "Poe versus Dickens: An Ambiguous Relationship." In *A Descent into Edgar Allan Poe and His Works: The Bicentennial*, edited by Beatriz González Moreno and Margarita Rigal Aragón, 3–24. Oxford: Peter Lang, 2010.

Garland-Thomson, Rosemarie. *Extraordinary Bodies: Figuring Physical Disability in American Culture and Literature*. New York: Columbia University Press, 1997.

Goldney, Grace. *Marion's Choice. Aunt Judy's Magazine* 8, no. 112–14 (1870): n. pag.

Granshaw, Lindsay. "'Upon This Principle I Have Based a Practice': The Development and Reception of Antisepsis in Britain, 1807–1890." In *Medical Innovations in Historical Perspective*, edited by John V. Pickstone, 17–46. Basingstoke: Palgrave, 1992.

Griffith, D. W., dir. *The Wooden Leg.* 1909; [New York City?]: Biograph Company.

G. W. C. "Good Advice and a Wooden Leg." *Hearth and Home: An Illustrated Weekly Journal for Gentlewomen*, no. 116 (1893): 419.

Haggard, H. Rider. *King Solomon's Mines.* Project Gutenberg. Last modified October 12, 2012. https://www.gutenberg.org/files/2166/2166-h/2166-h.htm.

Haley, Bruce. *The Healthy Body and Victorian Culture.* Cambridge: Harvard University Press, 1978.

Hamilton, David. "The Nineteenth-Century Surgical Revolution—Antisepsis or Better Nutrition?" *Bulletin of the History of Medicine* 56, no. 1 (1982): 30–40.

Handley, Neil. "Artificial Eyes and the Artificialization of the Human Face." In *Devices and Designs: Medical Technologies in Historical Perspective*, edited by Carsten Timmermann and Julie Anderson, 97–111. Basingstoke: Palgrave, 2006.

Hardy, Thomas. *The Woodlanders.* Oxford: Oxford University Press, 2009.

———. *Jude the Obscure.* Oxford: Oxford University Press, 2008.

Hasegawa, Guy R. *Mending Broken Soldiers: The Union and Confederate Programs to Supply Artificial Limbs.* Carbondale: Southern Illinois University Press, 2012.

Hazlitt, William Carew. *The New London Jest Book.* London: Reeves and Turner, 1871.

"He Fixed Her with His Glassy Eye". *Cheshire Observer*, April 5, 1879.

"Her Fatal Sneeze". *Illustrated Chips*, no. 2 (1890): 26.

Herschbach, Lisa. "Prosthetic Reconstruction: Making the Industry, Re-Making the Body, Modelling the Nation." *History Workshop Journal* 44 (1997): 23–57.

Holmes, Martha Stoddard. *Fictions of Affliction: Physical Disability in Victorian Culture.* Ann Arbor: University of Michigan Press, 2004.

Hood, Thomas. *Miss Kilmansegg and Her Precious Leg.* In *Selected Poems of Thomas Hood*, edited by John Clubbe, 193–270. Cambridge: Harvard University Press, 1970.

Horn, Pamela. *Pleasures and Pastimes in Victorian Britain.* Stroud: Amberley Publishing, 1999.

[Horne, Richard H.]. "Dust; or Ugliness Redeemed." *Household Words* 1, no. 16 (1850): 379–84.

Jain, Sarah S. "The Prosthetic Imagination: Enabling and Disabling the Prosthesis Trope." *Science, Technology, & Human Values* 24, no. 1 (1999): 31–54.

Jerrold, William Blanchard. "Eyes Made to Order." *Household Words* 4, no. 81 (1851): 64–66.

Jones, Claire L., ed. *The Medical Trade Catalogue, 1870–1914.* London: Pickering & Chatto, 2014.
———.. *Rethinking Modern Prostheses in Anglo-American Commodity Cultures, 1820–1939.* Manchester: Manchester University Press, 2017.
Ketabgian, Tamara. *The Lives of Machines: The Industrial Imaginary in Victorian Literature and Culture.* Ann Arbor: University of Michigan Press, 2011.
Kirkup, John. *A History of Amputations.* London: Springer-Verlag, 2007.
Koven, Seth. "Remembering and Dismemberment: Crippled Children, Wounded Soldiers, and the Great War in Great Britain." *The American Historical Review* 99, no. 4 (1994): 1167–1202.
Kurzman, Steven L. "Presence and Prosthesis: A Response to Nelson and Wright." *Cultural Anthropology* 16, no. 3 (2001): 374–87.
Lawrence, Christopher, and Richard Dixey. "Practising on Principle: Joseph Lister and the Germ Theories of Disease." In *Medical Theory, Surgical Practice: Studies in the History of Surgery*, edited by Christopher Lawrence, 153–215. London: Routledge, 1992.
"Legs: Wooden and Otherwise". *All the Year Round* 14, no. 350 (1875): 463–67.
Lieffers, Caroline. "Itinerant Manipulators and Public Benefactors: Artificial Limb Patents, Medical Professionalism and the Moral Economy in Antebellum America." In *Rethinking Modern Prostheses in Anglo-American Commodity Cultures, 1820–1939*, edited by Claire L. Jones, 137–57. Manchester: Manchester University Press, 2017.
[Lockhart, T.] "Prince Rupert's Emerald Ring." *Chambers's Journal of Popular Literature, Science and Arts* 12, no. 593 (1895): 300–4.
Lubin, Sigmund, dir. *A Cork Leg Legacy.* 1909; [Philadelphia?]: Lubin Manufacturing Company, 1909.
"The Man with the Glass Eye". *Pick-Me-Up*, no. 83 (1890): 78.
Mayhew, Henry. *London Labour and the London Poor: A Cyclopaedia of the Condition and Earnings of Those That Will Work, Those That Cannot Work, and Those That Will Not Work.* Vol. 1. London: Griffin, Bohn, and Company, 1861.
McKean, D. B. "A Wig and a Wooden Leg." *Chatterbox*, no. 4 (1886): 27.
Melville, Herman. *Moby-Dick.* London: Penguin, 2012.
Mitchell, David T., and Sharon L. Snyder. *Narrative Prosthesis: Disability and the Dependencies of Discourse.* Ann Arbor: University of Michigan Press, 2000.
Morgentaler, Goldie. "Dickens and the Scattered Identity of Silas Wegg." *Dickens Quarterly* 22, no. 2 (2005): 92–101.
[Morley, Henry]. "Ground in the Mill." *Household Words* 9, no. 213 (1854): 224–27.
"Mr. Aked Denounces Wigs". *The Penny Illustrated Paper*, July 27, 1907.
"Mr. Wegg and His Class". *All the Year Round* 18, no. 441 (1877): 250–53.
Nevett, Terence. "Advertising." In *Victorian Periodicals and Victorian Society*, edited by J. Don Vann and Rosemary T. Van Arsdel, 219–34. Toronto: University of Toronto, 1994.

O'Connor, Erin. "'Fractions of Men': Engendering Amputation in Victorian Culture." *Comparative Studies in Society and History* 39, no. 4 (1997): 742–77.

———. *Raw Material: Producing Pathology in Victorian Culture.* Durham: Duke University Press, 2000.

Ofek, Galia. *Representations of Hair in Victorian Literature and Culture.* Aldershot: Ashgate, 2009.

Otis, Laura. *Literature and Science in the Nineteenth Century: An Anthology.* Oxford: Oxford University Press, 2002.

Ott, Katherine. "Carnage Remembered: Prosthetics in the US Military since the 1860s." In *Materializing the Military*, edited by Bernard Finn and Barton C. Hacker, 47–64. London: Cromwell, 2005.

———. "Hard Wear and Soft Tissue: Craft and Commerce in Artificial Eyes." In *Artificial Parts, Practical Lives: Modern Histories of Prosthetics*, edited by Katherine Ott, David Serlin, and Stephen Mihm, 147–70. New York: New York University Press, 2002.

———. "The Sum of Its Parts: An Introduction to Modern Prosthetics." In *Artificial Parts, Practical Lives: Modern Histories of Prosthetics*, edited by Katherine Ott, David Serlin, and Stephen Mihm, 1–44. New York: New York University Press, 2002.

Paget, Henry. *One Leg: The Life and Letters of Henry William Paget, First Marquess of Anglesey, K.G. 1768–1854.* London: Leo Cooper, 1996.

[Parker, Frances], the Countess of Morley. *The Flying Burgermaster: A Legend of the Black Forest.* n.p.: F. Morley, 1832.

"Peg-Legged Bob". *All the Year Round* 11, no. 280 (1874): 563–68.

Pernick, Martin S. *A Calculus of Suffering: Pain, Professionalism and Anesthesia in Nineteenth Century America.* New York: Columbia University Press, 1985.

Phillips, Gordon. *Best Foot Forward: Chas. A. Blatchford & Sons Ltd. (Artificial Limb Specialists) 1890–1990.* Cambridge: Granta, 1990.

Poe, Edgar Allan. "The Man That Was Used Up: A Tale of the Late Bugaboo and Kickapoo Campaign." In *The Works of the Late Edgar Allan Poe*, by Edgar Allan Poe. Vol. 4: 315–25. New York: Blakeman & Mason, 1859.

———. "The Spectacles." In *Prose Tales (Second Series) Arthur Gordon Pym*, by Edgar Allan Poe, 161–98. Boston: Dana Estes, 1884.

ProQuest. *British Periodicals.* Accessed July 5, 2018. https://www.proquest.com/products-services/british_periodicals.html.

Quayson, Ato. *Aesthetic Nervousness: Disability and the Crisis of Representation.* New York: Columbia University Press, 2007.

Richards, Thomas. *The Commodity Culture of Victorian England: Advertising and Spectacle, 1851–1914.* Stanford: Stanford University Press, 1990.

"A Romance of War". *Kind Words*, May 1, 1872: 145.

Rosenburg, R. B. "'Empty Sleeves and Wooden Pegs': Disabled Confederate Veterans in Image and Reality." In *Disabled Veterans in History*, edited by David A. Gerber, 204–28. Ann Arbor: University of Michigan Press, 2012.

Rowland, Alexander. *The Human Hair, Popularly and Physiologically Considered with Special Reference to Its Preservation, Improvement and Adornment, and the Various Modes of Its Decoration in All Countries*. London: Piper, Brothers & Co., 1853.

Schaffer, Talia. *Romance's Rival: Familiar Marriage in Victorian Fiction*. Oxford: Oxford University Press, 2016.

Scott, J. Crawford. "The Flaxen Wig." *The Gentleman's Magazine* 266 (1889): 105–16.

Siebers, Tobin. "Disability as Masquerade." *Literature and Medicine* 23, no. 1 (2004): 1–22.

Slavishak, Edward Steven. "Artificial Limbs and Industrial Workers' Bodies in Turn-of-the-Century Pittsburgh." *Journal of Social History* 37, no. 2 (2003): 365–88.

Sobchack, Vivian. "A Leg to Stand On: Prosthetics, Metaphor, and Materiality." In *The Prosthetic Impulse: From a Posthuman Present to a Biocultural Future*, edited by Marquard Smith and Joanne Morra, 17–41. London: MIT Press, 2006.

Stainthorp, Clare. "Activity and Passivity: Class and Gender in the Case of the Artificial Hand." *Victorian Literature and Culture* 45, no. 1 (2017): 1–16.

Stark, James F. "'Recharge My Exhausted Batteries': Overbeck's Rejuvenator, Patenting, and Public Medical Consumers, 1924–37." *Medical History* 58, no. 4 (2014): 498–518.

"The Stricken Fawn". *Fun* 58, no. 1489 (1893): 219.

Sussman, Herbert, and Gerhard Joseph. "Prefiguring the Posthuman: Dickens and Prosthesis." *Victorian Literature and Culture* 32, no. 2 (2004): 617–28.

Sutherland, John. *Victorian Novelists and Publishers*. London: Athlone Press, 1976.

Sweet, Ryan. "'Get the Best Article in the Market': Prostheses for Women in Nineteenth-Century Literature and Commerce." In *Rethinking Modern Histories of Prostheses in Anglo-American Commodity Cultures, 1850–1960*, edited by Claire L. Jones, 114–36. Manchester: Manchester University Press, 2017.

———. "Pirates and Prosthetics: Manly Messages for Managing Limb Loss in Victorian and Edwardian Adventure Narratives." In *The Victorian Male Body: The Diverse Embodiment of White Masculinity in the Nineteenth Century*, edited by Joanne Parsons and Ruth Heholt, 87–107. Edinburgh: Edinburgh University Press, 2018.

Trollope, Anthony. *The Bertrams*. Leipzig: Bernhard Tauchnitz, 1859.

von Herkomer, Hubert. *The Last Muster*. 1875. Oil on canvas. 214.5 × 159 cm. Lady Lever Art Gallery, Liverpool.

Warne, Vanessa. "'If You Should Ever Want an Arm': Disability and Dependency in Edgar Allan Poe's "The Man That Was Used Up"'. *Atenea* 25, no. 1 (2005): 95–105.

———. "Artificial Leg." *Victorian Review* 34, no. 1 (2008): 29–33.

———. "'To Invest a Cripple with Peculiar Interest': Artificial Legs and Upper-Class Amputees at Mid-Century." *Victorian Review* 35, no. 2 (2009): 83–100.

"Wigs". *All the Year Round* 9, no. 225 (1873): 448–52.

Wolf, Jacqueline H. *Deliver Me from Pain: Anesthesia and Birth in America*. Baltimore: Johns Hopkins University Press, 2011.

Woodforde, John. *The Strange Story of False Teeth*. London: Routledge and Kegan Paul, 1968.

———. *The Strange Story of False Hair*. London: Routledge and Kegan Paul, 1971.

Wynter, Andrew. "The Artificial Man." *Once a Week* 1, no. 11 (1859): 218–20.

Youngquist, Paul. *Monstrosities: Bodies and British Romanticism*. Minneapolis: Minnesota University Press, 2003.

Yuan, David D. "Disfigurement and Reconstruction in Oliver Wendell Holmes's 'The Human Wheel, Its Spokes and Felloes.'" In *The Body and Physical Difference: Discourses of Disability*, edited by David T. Mitchell and Sharon L. Snyder, 71–88. Ann Arbor: University of Michigan Press.

Zemka, Sue. "1822, 1845, 1869, 1893, and 1917: Artificial Hands." *BRANCH: Britain, Representation and Nineteenth-Century History*, edited by Dino Franco Felluga [2015]. http://www.branchcollective.org/?ps_articles=sue-zemka-1822-1845-1869-1893-and-1917-artificial-hands.

CHAPTER 2

Constructing and Complicating Physical Wholeness

I think you had better not go to look at him. He's a dreadful object—the worst I've seen. They cut off his legs close to the trunk, his arms at the shoulders, the nose and ears. He was such a handsome fellow, too! But I tell you, sir, now he's nothing better than a human bundle—a lump of breathing, useless flesh.

—Ernest G. Henham, "A Human Bundle," *Temple Bar* 111, no. 438 (1897): 57–58.

Published in the metropolitan middle-class family magazine *Temple Bar* in 1897, Canadian-British author Ernest G. Henham's short story "A Human Bundle" is a literary source that in hyperbolic terms perpetuates fears about physical loss, anxieties central to nineteenth-century Western bodily discourse.[1] The quotation above, from a horrified medical student who has witnessed the shocking amputation of an unfortunate young man's legs, arms, nose, and ears, reveals what the loss of body parts meant ontologically in the nineteenth century. For the medical student, the patient is neither human nor useful but rather "nothing better than a human bundle."[2] The student's harsh assessment is partly justified by the egregious nature of the medical procedures undertaken, but such raises the question, how was bodily loss viewed more generally in the nineteenth century? If the patient's body in this story is incomplete, then what did a whole body look like? What constituted a physical loss? If the losses that

R. Sweet, *Prosthetic Body Parts in Nineteenth-Century Literature and Culture*, Palgrave Studies in Nineteenth-Century Writing and Culture, https://doi.org/10.1007/978-3-030-78589-5_2

Henham's character experience render him less than human and "useless," then what did it mean to lose or not be born with a leg, an eye, or a head of hair in this period? What historical factors underpin a privileging of physical wholeness and a fear of incompleteness?

This chapter explores how, as the concept of physical normalcy became increasingly reinforced as culturally dominant (the Self), those who were seen as missing body parts were marginalized (i.e. rendered Other). Buttressed by a post-Enlightenment belief that medicine and the emerging sciences could fix the issue of bodily loss, prostheses came to the fore as devices that could supposedly standardize aberrant bodies, making them aesthetically acceptable and useful. However, because prostheses were and remain devices that undermine binaries of Self/Other, organic/artificial, real/fake, and disabled/nondisabled, such devices also complicated the hegemony of organic wholeness. Their very production was mandated by preferences for physical completeness, but their implementation shifted definitions of what it meant to be whole. The conceptual complexity of the prosthetic provided material to fiction writers who responded to the growing dominance of physical wholeness. As I argue in the chapters following this one, fictional representations of prostheses simultaneously reinforced and complicated the hegemony of physical completeness. Such stories perpetuated fears of physical disaggregation while also bringing into question the very impulse to prostheticize.

The history of human attitudes to physical difference has garnered increased interest in recent years as disability studies has infiltrated mainstream scholarship. Lennard J. Davis's seminal book *Enforcing Normalcy* (1995) shows us how the concept of normalcy, against which disability is often understood, was constructed in the age of industrialization following the rise of bodily statistics and the corollary concept of the average man.[3] As I mention above, Rosemarie Garland-Thomson, who draws from Davis's influential work, coins the term "normate": a word that "names the veiled subject position of cultural self, the figure outlined by the array of deviant others whose marked bodies shore up the normate's boundaries."[4] Though not a disability-studies scholar per se, Sander Gilman shows us that images of disease invoke a process of boundary construction between the "healthy" observer, physician, or layperson, and "diseased" patient. Such a process, according to Gilman, helps to "localize" and "domesticate" disease, dissipating the fears of "collapse" that illness threatens.[5] Binary logics and social-constructivist frameworks link these critics' works. They show us that physical difference is not an inherent problem but is articulated as such in societies that construct, privilege, and perpetuate normalcy. The following contributes to debates surrounding the social-construction model of disability. I show that physical wholeness was a major constituent of normalcy articulated in ways that buttressed the power of the normate.

This chapter surveys the historical factors underpinning the rise of the normate. Beginning with an outline of various scientific theories that eroded Cartesian boundaries between body and mind, what follows parses legal and social changes, influential literary works, and prosthesis commerce. As I argue, each of these contexts contributed to the growth of physical normalcy and the denigration of difference.

Cultivating Completeness

The history of Western society's privileging of physical normalcy over difference is a contested one. Critics including Davis and Katherine J. Kudlick argue that this is a post-Enlightenment phenomenon.[6] Davis, for instance, asserts, "disability was not an operative category before the eighteenth century."[7] However, others, such as Quayson and Anolik contend that the privileging of normalcy is more longstanding. Anolik states, "The impulse of Western culture to define the human norm by the physical ideal and to construe the non-normative as dangerously close to the non-human actually predates the Enlightenment by millennia, as does the tendency to prioritize the norm and ignore the non-normative."[8] Elsewhere, Quayson writes, "as can be shown from an examination of folktales from all over the world, the plot of physical and/or social deformation is actually one of the commonest starting points of most story plots."[9] Given the recent work in this area, it seems probable that equivocal concepts to our modern notion of normalcy have existed for much longer than Davis acknowledges, but this does not diminish the value of his thesis. Indeed, what cannot be disputed is that the scientific, medical, and legal discourses of the eighteenth and nineteenth centuries buttressed, and sought to legitimize, boundaries between the normate and the physically aberrant. In this respect, the period of Western industrialization was without doubt a seminal moment in terms of constructing our modern understandings of the typical and the different body.

Theories of Body and Mind

Disability studies has shown us that the concept of the normal body is a social construction. As Davis states, "the idea of a norm is less a condition of human nature than it is a feature of a certain kind of society."[10] Among the most important and well-covered factors contributing to the rise of normalcy in the nineteenth century was the development of bodily statistics. Davis and Craton provide useful commentaries on this topic, but it is worth providing a summary here to illustrate how wholeness was constructed as a major constituent of the emerging discourse surrounding the "average" man.[11] Belgian mathematician, statistician, astronomer, and sociologist Lambert Adolphe Quetelet came up with the concept of "*l'homme moyen*," or "the average man," in his influential 1835 work *A Treatise on Man and the Development of His Faculties.*[12] As Craton summarizes, "Quetelet calculated the mathematical norms for a range of physical and social categories[,] everything from head circumference to age of marriage to criminal tendency[,] in order to draw a detailed portrait of the human norm."[13] For Quetelet, his average man constituted a kind of paradoxical ideal: "an individual who epitomized in himself … all the qualities of an average man, would represent at once all the greatness, beauty and goodness of that being."[14] Quetelet's notion of *l'homme moyen* became popular in England in the 1830s. It was well received by intellectuals and stimulated the development of the contemporary disciplines of social science.[15] The implications of Quetelet's work on non-normative bodies are concisely noted by Davis: "When we think of bodies, in a society when the concept of the norm is operative, then people with disabilities will be thought of as deviants."[16] Intellectually, Quetelet's theories set the foundations for eugenics, a set of ideas that sought to eradicate physical and mental differences seen as negative.

While the concept of physical wholeness is not evoked directly in Quetelet's concept of *l'homme moyen*, his average man has two eyes, two arms, two legs, two arms, and so on. In other words, he is physically complete. Associating health with the average and illness/deformity with those who deviate from it, Quetelet observes:

> If the average man were completely determined, we might, as I have already observed, consider him as the type of perfection; and every thing differing from his proportions or condition, would constitute deformity and disease;

> everything found dissimilar, not only as regarded proportion and form, but as exceeding the observed limits, would constitute a monstrosity.[17]

Such a view provides evidence for Bruce Haley's observations that Victorian understandings of health centred on the concept of completeness. Haley suggests that "[n]o topic more occupied the Victorian mind than Health," and that "wholeness," along with functionality and vitality, was a key component of what constituted health.[18] As Haley states, "health is a state of functional and structural wholeness. In an organism the two are related, for a structure becomes functional when viewed as part of a living whole."[19] An implication of such a view of health was that those who exhibited even the slightest degree of incompleteness were rendered unhealthy—a label with grave consequences.

Another significant context that contributed to the stigmatization of physical incompleteness, and deviance more widely—which intersected with Quetelet's work in terms of the classification of physical difference—was the popularity of the science of physiognomy. The philosophy that one can ascertain the moral qualities of another by assessing her/his appearance, in particular her/his face, has existed since classical times.[20] But such a system was codified and popularized in the late eighteenth and early-to-mid nineteenth centuries by Swiss pastor Johann Kaspar Lavater, whose influential *Essays on Physiognomy* were first published in German between 1775 and 1778.[21] As a result of rapid population growth, urbanization, and changing social dynamics, physiognomy became an important tool in the nineteenth century: "physiognomy provided a new code for conformity as people became concerned not just with judging others, but with how others were judging them."[22] Thomas Holcroft's cheap English translation of Lavater's *Essays on Physiognomy* was reissued eighteen times from the late 1780s through to the late 1860s alone, indicating the popularity of physiognomy in Victorian England.[23] John Graham notes that Lavater's essays were published 156 times before 1940.[24]

Regarding physical wholeness, Lavater's writings and the subsequent works that drew upon his philosophies reinforced prejudices against physical loss and asymmetry developed in the Bible. For instance, in 1805 Richard Payne Knight, a reputed British arbiter of taste, commented on the peculiar but nonetheless "naturalized" prejudice against animals that have non-normative facial features: "if we were to meet with a beast with one eye … we should without inquiry, decide it to be a monster, and turn from it with abhorrence[.] … [T]he Creator having formed the one regular, and the other irregular, we habitually associate ideas of regularity to the perfection of the one, and ideas of irregularity to the perfection of the other."[25] This passage recalls the biblical correlation between ocular aberrance and moral corruption. For instance, in the Old Testament it is asserted that to "the idol shepherd that leaveth the flock[,] ... his arm shall be clean dried up, and his right eye shall be utterly darkened."[26] Here the loss of an eye is seen as an apt punishment for or mark to distinguish negligent character. A similar sentiment is echoed in the New Testament gospel of Luke: "when thine eye is single [good, healthy, or clear], thy whole body also is full of light; but when *thine* eye is evil [wicked, bad, or diseased] thy whole body also *is* full of darkness."[27] The latter passage was quoted and reiterated by Lavater, who added, "[i]t is as physiognomonically [*sic*] true, also, that when nothing is oblique, sinister, dark, rough, incongruous, heterogeneous, in the body, then is all health and harmony, and every object bright."[28] Despite Lavater's use of this quote as support for his physiognomical treatise, such a stigmatized view of having one eye is not entirely consistent in the Bible. There is a famous verse in the Gospels where Christ exhorts his followers to gouge out an eye if it causes them to sin, concluding with "it is better for thee to enter into life with one eye, rather than having two eyes to be cast into hell fire."[29] Here the verse aligns having one eye with purity or virtue, which is significant as Jesus is arguably overturning the injunction in Leviticus that prohibits the blind (and otherwise impaired) from entering the Temple.[30] Lavater aligned his view on eyes with George Daumer, who in 1702 claimed: "The eye appertains more to the soul than any other organ."[31] It is suggested that to display two healthy eyes is to display good character. Though he ambiguously distances his readings of eyes with those quoted, Lavater also cites Paracelsus, who applied the following traits to "Small, and deep sunken eyes ... bold in *opposition*": "not discouraged, intriguing, and active in wickedness; capable of suffering much" (emphasis added).[32] For Lavater and his followers, those who deviated from the standard, conspicuously whole, body created by God manifested visibly an aberration that ran much deeper. Such views placed additional pressure on the notion of self-presentation. Those with physical deformities were forced to conceal their differences to avoid negatives physiognomic assessments. To use current terminology, they were encouraged to pass.[33]

In a manner not dissimilar to the champions of Lavater and the proponents of the related science of phrenology, such as Franz Joseph Gall, who linked physical features (of the human cranium in particular) to behavioural traits, many scientists, philosophers, and physicians continued to controversially erode Cartesian boundaries between mind and body. The impact of this erosion further solidified the premium on physical completeness. From the 1830s onwards, a physiological model for understanding the human form emerged, which often emphasized the reciprocal rather than the previously assumed independent natures of the mind and the body. Studies of the body, the physiology of the brain, and the nervous system thereby took on increased significance. Stressing the importance of the body in 1835, popular English physician and sanitary reformer Thomas Southwood Smith demanded that physiology should be the basis of all study of humankind:

> The mind is dependent on the body: hence an acquaintance with the physiology of the body should precede the study of the physiology of the mind. The constitution of the mind must be understood before its powers and affections can be properly developed and directed, hence a knowledge of the physiology of the mind is essential to a sound view of education and morals.[34]

Alexander Bain's *The Senses and the Intellect* (1855) revived attention in the link between mind and body.[35] As he suggested, "[a]lthough Subject and Object (Mind and Matter) are the most diametrically opposite facts of our experience … there is a concomitance or connection between mind and a material organism."[36] He would later reaffirm:

> Instead of supposing that mind is something indefinite, elastic, inexhaustible,—a sort of perpetual motion, or magician's bottle, all expenditure, and no supply, we now find that every single throb of pleasure, every smart of pain, every purpose, thought, argument, imagination, must have its fixed quota of oxygen, carbon, and other materials, combined and transformed in certain physical organs.[37]

Contemporaries who supported the links made between mind and body included evolutionary biologist Herbert Spencer, German physiologist and philosopher Wilhelm Wundt, the anatomist T. H. Huxley, and the mathematician and philosopher W. K. Clifford. Together, these figures contributed to a new tradition in the sciences that we might usefully call "psycho-physiology." While, as Lorraine J. Daston, Rick Rylance, and Roger Smith have observed, all of these proponents trod a fine line between attempting to create a characteristically physiological discipline for analysing the mind and downplaying accusations of materialism, implicit in each account was an attempt to collapse Cartesian dualism.[38] William A. Cohen pays testament to the cultural influence of such work in his monograph *Embodied (2008)*. Here he goes so far as to claim that "embodiment came to be the untranscendable horizon of the human."[39]

In terms of the relationship between physical difference and mental state, pre-eminent psychiatrist Henry Maudsley, another key figure in the psycho-physiology movement, bolstered the links between body and mind carved out by Lavater in his 1874 work *Responsibility in Mental Disease*:

> There is not an organ in the body which is not in intimate relation with the brain by means of its paths of nervous communication, which has not, so to speak, a special correspondence with it through internuncial fibres, and which does not, therefore, affect more or less plainly and specially its function as an organ of mind. It is not merely that a palpitating heart may cause anxiety and apprehension, or a disordered liver gloomy feelings, but there are good reasons to believe that each organ has its specific influence on the constitution and function of mind[.][40]

For Maudsley, the brain was not the only physiological matter that could influence the mind; for him, other parts of the body could affect the temperament of a subject in many ways. Associating physical deviances with specific moods and behaviours, Maudsley implied that those who display physical difference are often mentally aberrant therefore legitimizing the work of Quetelet, who claimed that the normal body was the healthiest in all regards. Most severely, Maudsley asserts, "[m]ultitudes of individuals come into this world weighed with a destiny against which they have neither the will, nor the power to contend; they are the step-children of nature and groan under the worst of all tyrannies—the tyranny of a bad organisation."[41] Endorsing a theory that suggested physical and mental degeneration are passed on by a process of atavism, Maudsley supports a deterministic understanding of the human condition that stigmatized physical difference.[42]

Criminal anthropologist Cesare Lombroso "celebrated the penetration of a quantifying materialism."[43] He buttressed the links forged by the psycho-physiologists between mind and body as he sought to demonstrate empirically a correlation between "monstrousness" and criminality. As David G. Horn explains:

> For Lombroso, what truly distinguished the modern era from all that had come before was the triumph of "the number" over the "vague opinions, prejudices, and vain theories" that had circulated from the folk to the learned community and back again. Among these was the conviction—shared by the masses and the greatest physiologists and psychologists—that there was an "immeasurable abyss" between the world of "life and intelligence" and the world of "brute materiality." But numbers, he observed in a volume on weather and mental illness, had been able to bridge the abyss, entering "with their marvellous power" even into the "mysterious world of life and the intellect."[44]

The influences of physiognomy and phrenology on Lombroso's thinking are obvious and were recognized by the criminal anthropologist himself, but the impact of Quetelet's use of social statistics was even more profound.[45] Given the influences of these three popular concepts, it is unsurprising that Lombroso's work was so influential in the late nineteenth century even though his most famous book *Criminal Man* (1876) is still yet to be translated fully into English.[46] Britain and America were, after all, already primed for empirical investigation of the link between physical difference and criminal behaviour by both the popularity of sciences of physiognomy and phrenology and the prominence of physically aberrant criminal characters in literary works—as attested to in studies by David T. Mitchell and Sharon L. Snyder and by Quayson.[47] Neil Davie shows us that despite an explicit resistance to Lombroso's theories exhibited by many British criminologists, they were met with favourable reaction by a number of notable British writers on psychology, including not only Francis Galton and Havelock Ellis but also asylum-based psychiatrists on the margins of the criminal-justice system, including Thomas Clouston (of the Edinburgh Royal Asylum), Samuel Strahan (of the Northampton County Asylum), and Alfred F. Tredgold (author of the widely read textbook *Mental Deficiency* [1908]).[48] Though Lombroso's theories were certainly controversial, they were very much part of the social consciousness at the fin de siècle in both Britain and America. The deviant body became an increasingly centralized topic as anxieties about the physically aberrant grew. Such a process of marginalization reinforced the dominance of physical wholeness.

In addition to the feared criminal traits of those who displayed physical difference, the premium on physical integrity was also bolstered by medical and folkloric views that saw the aberrant body as a potential threat to normative society. Drawing on disability theorist Tobin Siebers's framework of the "evil eye event"—a dynamic in which "accusation exaggerates" physical and mental differences "until they take on a supernatural dimension"—Susan M. Schweik shows how late nineteenth-century unsightly begging laws sought to protect the normate, in particular women, from bearing witness to dismembered or deformed beggars.[49] Schweik reveals that fears proliferated stemming from a belief that seeing physically atypical people could have pathological effects on women, producing symptoms such as "seizures, hysteria, or 'conniption' [a fit of hysterics]."[50] Anxieties surrounding "maternal impression"—the theory that if a pregnant woman witnessed a person with a deformity, the "shock" caused by such an encounter could result in her unborn child bearing a similar "affliction"—also thrived in this period. As Jan Bondeson shows us, 170 articles on maternal impression appeared in US scientific journals between 1839 and 1920.[51] Underlining the intolerance of Western society to witnessing physical difference in real life, Schweik shows us how several American cities criminalized the display of perceived ugliness in public spaces during the late nineteenth century.[52] Though such abominable legislation did not appear in Britain in the same period, similar fears existed there. Martha Stoddard Holmes explains that amidst an environment in which many conflicting theories arose, "any physical impairment had the potential to be perceived as transmissible by contact; by miasmic air; by a combination of contact, environment, and individual constitution; or perhaps simply by the social class into which one was born."[53] The sheer variety of explanations listed by Holmes reveals the level of anxiety that surrounded the supposed risk of becoming physically disabled in Victorian Britain.

Legal and Social Factors

Linked to Quetelet's drive for standardization, another event that contributed much to anxieties surrounding physical difference was the 1834 Poor Law Amendment Act. The Amendment Act introduced a centralized system to manage administering relief to the poor. The new system was brought about, at least in part, in response to fears from social reformers that the decentralized parish-based system of poor relief implemented by the Elizabethan poor laws was inconsistent and, at times, misused. The new system sent able-bodied men seeking relief to the workhouse while providing limited out-relief to those deemed deserving—namely, those unable to work, including the young, the elderly, and the disabled. Despite this tightening of the law, as Holmes reveals, fears of abuse remained. Much discussion about the extent to which the unwaged disabled were deserving of relief perpetuated as a result.[54] This context is discussed in more detail in Chap. 4, which considers anxieties about physical completeness alongside social mobility, but it is worth noting at this point that the act brought public attention to the ability of aberrant bodies, the classifying of such bodies, and an association of physical difference with mendicancy.

Changing meanings of work further exacerbated links between physical difference and an inability to work, a factor that had severe implications for men—the primary breadwinners in this period. Works such as Thomas Carlyle's *Past and Present* (1843) and Samuel Smiles's *Self Help* (1859) were influential as they propounded the importance of industriousness and renounced idleness.[55] John Tosh plots the rise of this ideology between the Reform Acts of 1832 and 1884, when suspicions surrounding privilege gained momentum and faith in the idea of individual autonomy took its place.[56] Such an emphasis on autonomy and industry meant that those exhibiting physical difference were seen as lacking the necessary attributes to succeed in life. Erin O'Connor explains that "Victorian ideals of health, particularly of male health, centered on the concept of physical wholeness: a strong, vigorous body was a primary signifier of manliness, at once testifying to the existence of a correspondingly strong spirit and providing that spirit with a vital means of material expression."[57] As we learn from Henham's "The Human Bundle," where the maimed man is branded "a lump of breathing, useless flesh," for men especially, physical difference was seen as an indicator of a subject's inability to work.[58] This was the case on both sides of the Atlantic. "Nowhere is the disabled figure more troubling to American ideology," writes Garland-Thomson, "than in relation to the concept of work."[59] Highlighting public curiosity surrounding the ability of atypical bodies to work, Garland-Thomson elsewhere observes that amputees performing the most commonplace of tasks became public spectacles: "the 'Armless' or 'Legless Wonders' who performed mundane tasks like sewing, writing, riding a bicycle, or drinking tea [in freak shows] were at once routine and amazing, both assuringly domestic and threateningly alien."[60] While the public were fascinated by and yet unsure of the working capacities of physically incomplete subjects, as O'Connor shows, contemporary prosthetists sought to cash in on prejudices by producing devices that could enable disabled men to return to work.[61]

While physically incomplete men were often believed to be unable to work, a point evidenced in autobiographical memoirs by disabled factory workers, such as *A Narrative of the Experience and Sufferings of William Dodd* (1841), women missing body parts were regularly represented as unmarriageable and thus not useful in terms of procreation or social obligations—they were seen as part of the "superfluous women" problem.[62] Exceptions to this trend were imagined in literary sources, such as Wilkie Collins's *Poor Miss Finch* (1872), a topic that Holmes discusses at length in *Fictions of Affliction* (2004).[63] But in reality, barriers existed that made it difficult for physically and cognitively different women to enter the marriage market. For instance, contemporary medical science sometimes linked physical aberrance to hereditary illness and deformity, raising concerns about women exhibiting physical losses entering the arena of marriage. I focus specifically on these contexts and how anxieties surrounding women with atypical bodies were exacerbated and interrogated by literary depictions in Chap. 5.

In addition to these concerns, the idea was also prevalent that physical wholeness comprised a full head of hair and a complete set of teeth. Galia Ofek notes how hair "became increasingly significant to the formation of self-image since the growth of the European city and the ascendance of the bourgeoisie in the early stages of mercantile capitalism." During the nineteenth century, "among the middle classes in Victorian England, women's hair turned into a salient focal point as fashion dictates and social mores prohibited bare hands, legs, and other parts which were covered for modesty's sake, thereby turning hair, neck, and shoulders into the 'focus of sexual interest' which substituted 'for all the rest.'"[64] This explains, to some extent, the significance of hair among women, but what about its significance for men? It is true that men's haircuts grew shorter and less ornate during the nineteenth century. But displaying a full head of hair was still important for men as baldness signified physical decline, something that they were keen to ward off. In Kay Heath's discussion of male-midlife anxieties, she analyses an advertisement for the Edwardian "hair grower" Tatcho in which maker George R. Sims challenges British men with the headline, "Bald, grey, or sparse of Hair: what are your chances in life?"[65] Quoting the advertisement, Heath writes, "baldness 'is a touchy subject with most men,' because age can be determined by hair rather than years: while a shiny pate make one 'old at thirty,' 'with a good head of hair' a man 'may look young at fifty.'" As Heath shows us, in a time when "concepts like occupation and physical prowess replaced traditional notions of privilege to determine the measure of a man," it became less desirable for men to display accepted signs of old age in public spaces.[66]

Meanwhile, the demand for a full set of clean, healthy teeth gained purchase as urbanization and the promises of capitalism encouraged individuals to "cultivate good surface impressions."[67] M. D. K. Bremner has noted that the demand for perfect teeth was first cultivated in modern society in early eighteenth-century France: "High society developed a pagan view of life. Beauty became the dominant note of the age. Well-to-do people sought health, comfort and good looks and tried above all to retain their youth. Since a toothless face can look neither young nor beautiful, there was great demand for dentistry."[68] Pioneering dentist Pierre Fauchard reaped the rewards of such a culture, but it was not until the 1840s that dentistry began to professionalize more widely across Europe and North America.[69] The rise of dentistry at this time shows the impact and prevalence of physiognomy in the early to mid-Victorian period. As Karen Halttunen notes, advice manuals published between 1830 and 1850 implored readers to take care of their personal appearances: "Surface impressions were essential to success in the world of strangers ... because appearances revealed character."[70] The desire for good teeth was exacerbated by the sheer volume of advertisements for teeth-related products that potential consumers in Britain and America were bombarded with—in particular following the Great Exhibition of 1851.[71] M. Cox et al. observe that "Patented tinctures and powders proliferated ... particularly in response to periodontal [gum] disease and [to aid] the removal of calculus [tartar, a hardened form of plaque]."[72] Peter A. Reichart meanwhile draws attention to the prevalence of the Areca nut in popular tooth pastes and powders.[73] Simulated by major developments in dentistry—including the introduction of anaesthesia and the use of vulcanite, a cheap and easy material to work with to manufacture false teeth bases—the mandate for good teeth was so great that by 1877 one journalist wrote:

> Natural teeth, clean, sound, and perfect, are essential to the comeliness of any human face. Defective teeth mar the handsomest features and cause us to turn away our gaze with a kind of disgust from a countenance otherwise faultlessly beautiful. Sound teeth not only add to the comfort and personal appearance, but contribute largely to the health of all, hence special and scrupulous attention should be paid to them daily, from early childhood, from the time when the first permanent tooth makes its appearance about the sixth year.[74]

To make a good first impression in a world where cultivating an appearance of physical wholeness was of growing importance, a good set of teeth became a fundamental requirement.

Literatures of Loss

Literary sources captured the ableist zeitgeist of the period while raising new anxieties concerning the neurological symptoms of physical loss. Ground in research on the intimacy of mind and body (especially the role of the nervous system as the joining matter between the two), the arrival of phantom limb syndrome into public discourse via literary fiction further exacerbated social fears of bodily loss. Coined in 1871 by the American neurologist and fiction writer Silas Weir Mitchell, phantom-limb syndrome describes the common sensation whereby, when a limb has been amputated, "the sufferer does not lose consciousness of its existence."[75] Mitchell's coinage stands to this day as the recognized term to describe this condition. According to Mitchell, phantom-limb sensations were so frequent that only five per cent of amputees did not experience them.[76] Mitchell suggested that this condition could manifest itself in several ways. It could cause symptoms such as a feeling that a lost limb was still attached and functioning; intense pain at an extremity no longer there; itchiness of a lost extremity; a feeling of "perpetual … automatic activity"; and a sensation of the "shortening of the absent member."[77] As Mitchell explains:

> Since the stump is the lowest *visible* point where pain or touch is felt, the sensorium or central organ of feeling gradually associates in place the lost hand or foot with the stump, the most remote existing part, impressions on which are referred to the lost limb. Hence arises a notion of shortening in the absent member—an idea which is more and more faintly contradicted by previous knowledge, and more and more reinforced by present subjective sensations.[78]

In other words, Mitchell suggests that the visual and haptic senses become confused because of nerve damage to the severed limb, leading to a sensation whereby an outer extremity becomes associated with the end of a stump.

Some of these symptoms were known to medical practitioners long before Mitchell's work on the topic. He noted himself that Ambroise Paré, a sixteenth-century pioneer of amputation and prosthetics, referred to these kinds of sensation in his work on limb loss.[79] But it was Mitchell's own fictional account of a quadruple amputee in 1866 that brought public attention to this medical topic. Prior to his coinage, other nineteenth-century medical practitioners, like Wilhelm Gottfried Ploucquet, Jeremais David Reuss, François Xanvier Swediaur, Gabriel Valentin, Thomas Young, Godofredus Theodatus Rhone, Johannes Müller, and Charles Bell, all wrote about phantom limbs, but none captured the public imagination in the same way as Mitchell.[80] The latter used literature as a tool to communicate his ideas to a wider audience, capitalizing on the popularity of the Gothic mode, while exploiting and further developing social anxieties regarding the loss of limbs.

Published anonymously, Mitchell's "The Case of George Dedlow" (1866), a sketch about a man who lost all four of his limbs during the American Civil War, humorously explored some of the neurological, psychological, and ontological implications of amputation.[81] Much to Mitchell's amazement, this sketch, which poses itself as a kind of medical report, was considered factual by many readers. According to Mitchell, "Inquiries were made as to the whereabouts of the sufferer, and in an interior county of New York a subscription was actually started for the unhappy victim."[82] While Mitchell was careful to distance himself from this short story by providing a corrective in his later article "Phantom Limbs" (1871), there were significant parallels between the symptoms described by the fictional George Dedlow and Mitchell's medical explanation. Though Dedlow does not define the condition that he describes as "phantom-limb" syndrome in Mitchell's literary work, his observations regarding the limb consciousness that he and the other inpatients at the Stump Hospital (Nashville, Tennessee) experience, in combination with the various sensations described—itching, pains, and cramps—bear uncanny resemblance to those parsed in the author's later medical piece.[83] In a slightly more figurative way than in "Phantom Limbs," using Dedlow as a mouthpiece, Mitchell ventriloquially explains the neurological phenomenon of phantom limbs in layman's terms, using a simile of a bell-wire to explain why a sensation of a lost limb remains after it has been removed. As Dedlow suggests, a severed nerve present in a stump is a like a bell-wire because it can be stimulated at any part of its course and still produce a signal at its extremity. Mitchell makes the same point in "Phantom Limbs," using the more relatable and bodily example of the sensation experienced when we hurt our "crazy-bone" (now known in the United Kingdom and the United States by the colloquial term the "funny-bone").[84] While

Mitchell's essay explained phantom-limb pain in a more intricate, medicalized way, the two pieces described many of the same neurological symptoms of limb loss. In fact, the only aspects arising from "George Dedlow" that Mitchell really refuted in his article were: first, the possibility of surviving quadruple amputation; second, the fact that phantom-limb pain only occurs in poorly formed stumps; and, third, that such extreme psychological states can be produced by amputations—Dedlow experiences delight while believing that he is united with his missing limbs at a spiritualistic séance before slipping into a dreary state of unhappiness.

The very symptoms of phantom-limb pain described by Mitchell reaffirmed beliefs in the importance of embodied wholeness. Mitchell's works revealed to relatively ignorant readers that losing a limb resulted not only in the loss of function of that part but also in severe and sometimes debilitating neurological symptoms. Losing a limb took on greater meaning as it represented a disruption of the overall structural unity of the body. By emphasizing the existence of muscular memory, a phenomenon that O'Connor suggests allowed artificial limbs to function, Mitchell's research into stump pathology suggested that phantom-limb pain was a product of the body's refusal to accept dismemberment, again bringing into focus both the nervous link between mind and body and the perceived importance of physical integrity.[85] However, complicating the premium placed on physical wholeness, stumps often had to be re-amputated to cure some amputees of severe phantom-limb symptoms. Mitchell described an example whereby this course of action was required in *Injuries of the Nerves* (1872).[86]

O'Connor draws our attention to the feminizing implications of the pathological stumps associated with phantom-limb disorder.[87] Revealing the heavily gendered view of contemporary medical practitioners, O'Connor quotes John Erichsen, who in his 1854 treatise *The Science and Art of Surgery* associated "pathologically constituted stumps" with "feminine susceptibility." According to Erichsen, most serious cases of intense stump pain and twitching arose "from constitutional causes, and invariably occurs in females, more particularly in those of the hysterical temperament, and who are subject to neuralgic pains elsewhere."[88] As O'Connor argues:

> The notion that stump pathology is not only feminine, but female, had far-reaching consequences for medical and social understandings of the male amputee, whose manhood was thereby implicated in an effeminate pain

> pattern. Altering the body in ways that were as psychically threatening as they were physically therapeutic, amputation raised the possibility that cutting off a man's limb could cut a man off from himself.[89]

O'Connor's concern is the male amputee. It is true that "jumpy stumps" presented a supposedly feminizing threat to male amputees, but the implications of such on female amputees should not be overlooked.[90] Jane Wood has noted how hysteria was a condition "whose clinical criteria could be modified to diagnose all the behaviours which did not fit the prescribed model of Victorian womanhood."[91] Female amputees were rendered even further from ideals of womanhood by neurotic stumps. The spasmodic and wild behaviour of such stumps exhibited the negative traits against which ideal womanhood was defined, providing an additional symptom of loss that female (as well as male) amputees needed to conceal to pass. The exposure of phantom-limb syndrome to a public audience therefore exacerbated fears of limb loss to both male and female readers.

In addition to works such as Erichsen's, which show that pathological stumps were part of medical consciousness before Mitchell's work on the topic, other literary writings, such as Frederick Marryat's *Jacob Faithful* (1834) and Charles Dickens's *Our Mutual Friend* (1864–1865), exhibited the curious and confounding behaviour of amputated stumps.[92] Though published before Mitchell's nuanced work on the topic, these stories fostered fears of bodily loss as they presented stumps as unruly and perplexing. Marryat's *Jacob Faithful*, a popular bildungsroman for children, features a character named Old Tom, a double-amputee naval veteran who lost his legs fighting in the battle of Trafalgar during the Napoleonic Wars. While Old Tom can be considered an archetypical good prosthesis-using navy veteran—alongside later fictional examples such as Dickens's Captain Cuttle from *Dombey and Son* (1846–1848) and Gruff and Glum from *Our Mutual Friend*—what is unique about Marryat's character is that he makes manifest not only the functional limitations of primitive prostheses, which were standard for Greenwich Hospital amputees, but also the nervous peculiarities that apparently so often accompanied limb loss.[93] Much to his son Tom's and the eponymous protagonist Jacob's surprise, Old Tom reveals that despite being amputated above both knees, his toes give him periodic discomfort. "[S]ometimes I feel them just as plain as if they were now on, instead of being long ago in some shark's maw," Old Tom reveals. "At nights I has the cramp in them till it almost makes me halloo out with pain. It's a hard thing when one has lost the sarvice of his legs, that all the feelings should remain. The doctor says as how its nervous."[94] Here, Old Tom reveals his own excruciating experience of what would later be labelled phantom-limb pain by Mitchell. Marryat presents Old Tom's curious, painful, and neurotic stumps as medical curios, evoking the nineteenth-century fascination with bodily abnormality exhibited in so-called freak shows.[95] Like these sideshows, Marryat's representation simultaneously exploited and contributed to contemporary anxieties related to physical difference; aligning with Gilman's observations, it utilized the dynamic of "healthy" observer versus pathologized observed.[96]

Dickens's depiction of Silas Wegg has been linked to Marryat's characterization of Old Tom. Michael Cotsell, for instance, highlights the similarities of the two characters in terms of how both alter the words of songs to fit the occasion.[97] The most significant likeness in terms of the cultural construction of wholeness, however, is that both characters display phantom-limb symptoms. While reading various accounts of misers to Mr. Boffin in Chap. 6 of Book 3 in *Our Mutual Friend*, the amputee Wegg becomes increasingly excited. His excitement is signified by the prodding and elevation of his wooden leg: he repeatedly "peg[s]" his comrade, Mr. Venus, and his leg "start[s] forward under the table, and slowly elevate[s] itself as he read[s]."[98] These impulsive movements can be read to display self-action on the part of his prosthesis, as both Herbert Sussman and Gerhard Joseph imply regarding Captain Cuttle in *Dombey and Son*, or just Wegg's comic absurdity.[99] And yet such neurotic behaviour could also be read as the result of a twitching stump, one plausibly affected by phantom-limb or jumpy-stump syndrome. In this way, we can read Dickens's representation in context with recent research that brings to light how his fiction described conditions before they had been medically recognized.[100] The involuntary movement of Wegg's leg eventually results in him losing balance and "dropp[ing] over sideways" onto Venus.[101]

The automatous behaviour of Wegg's peg, powered by what seems a pathologized stump, bears resemblance to the "curious spasmodic maladies" described by Mitchell in his later essay on phantom limbs.[102] Wegg is himself described as "spasmodic" when he attempts to pick himself up off the floor. The phallic resemblance of this unconscious behaviour—a body part becoming erect before reaching a violent climax and ending with the subject left in a "pecuniary swoon"—is undermined by what would have been seen as feminine, hysterical undertones.[103] His lack of self-control and compromised position at the end of this scene—where he lies on

Venus in a swoon-like state—emphasizes his apparently unmanned condition as an amputee.[104] Though troubling from a contemporary perspective, this scene was clearly intended to be comical. In fact, it was not uncommon for Dickens to use physical difference for comic purposes. For instance, in *Nicholas Nickleby* (1838–1839), the author encourages the reader to laugh at the villainous schoolmaster Wackford Squeers's mishaps that are brought about by his lack of full peripheral vision, a functional difference that arises from him having one eye.[105] Philip Hobsbaum writes that the "Jonsonian" humour of Squeers's representation makes tolerable what would otherwise seem an unbearable "rhetoric of indignation" towards Yorkshire schoolmasters.[106] In a similar fashion, Dickens uses grotesque comedy in *Our Mutual Friend* to lighten what might have otherwise appeared a frustrating story about a duplicitous "cripple" who almost upsets the status quo by climbing the social ladder (a narrative trajectory that I explore further in Chap. 4). Though at other times (even within the same narrative) much more sympathetic to issues of physical difference—think, for instance, of his representation of characters such as Esther Summerson, Joe Willet, and Captain Cuttle—in this instance, Dickens's depiction of Wegg's out-of-control stump exacerbates anxieties about losing limbs.[107] As we will see in Chap. 4, for the Dickens that wrote *Our Mutual Friend*, the amputee was at once fascinating, confounding, suspicious, and troubling. Regarding Dickens's depiction of stump maladies, however, like Marryat's earlier depiction and Mitchell's literary and medical writings, it represented (and exploited for fictional purposes) real-life issues facing amputees, thereby cultivating social fears of limb loss.

Persuasive Prosthetists

Before we turn to fictional representations of prostheses in the chapters that follow, it is important to consider the position the emerging prosthesis market assumed amidst the culture described so far. Aligning with trends in patent medicine that Thomas Richards, Claire L. Jones, and Jamie Stark have described, prosthetists of all types, including makers of limbs, eyes, teeth, and hair, capitalized on growing anxieties about the body and health, in particular the increasing mandate for physical completeness.[108] As O'Connor has revealed, the notion of rebuilding amputees to a condition of wholeness was commonly evoked in nineteenth-century prosthesis discourse, especially that involving male amputees.[109] The American prosthesis firm, A. A. Marks, one of the most famous and successful makers of its time, for instance, included the following testimony from the Atlanta *Christian Index and Southwestern Baptist* in its catalogue: "Mr. Marks has the most skilled mechanics in his manufactory, turning out frequently a dozen or more limbs a week. It is interesting to see his patrons leave their crutches in his office, and walk off apparently *whole*—men, too, who had lost both legs and who were brought in by attendants" (emphasis added).[110] After purchasing one of Marks's patented artificial legs with rubber foot, John McKenzie, a Civil War amputee, similarly testified, "I felt like a whole man again."[111] Limb makers such as Marks were keen to assert to potential users the abilities of their devices to recomplete supposedly disaggregated bodies. Testimony pages, such as the ones from which the quotations above are extracted, were important locations where makers could communicate the quality of their devices to potential consumers without attracting accusations of quackery.[112]

Artificial-limb makers were not the only ones to draw from the dominance of wholeness in their advertisements. French artificial-eye maker Auguste Boissonneau, who brought his highly rated enamel artificial eyes to Britain and Ireland in the early 1850s drew directly from many of the cultural factors described above that contributed to the stigmatization of those missing body parts. An advertisement printed in *Freeman's Journal and Daily Commercial Advertiser* in 1852 reads:

> Mr. Boissonneau has succeeded in removing the physiognomical defect, which is the consequence of such a loss, by his newly-devised Artificial Eyes, which patients can apply themselves with the greatest ease. The movements of the artificial substitute are so admirable that it is hardly possible to distinguish nature from art. The use of these eyes is fully appreciated by those who are aware how much irregularities of the face mar a man's career[.][113]

Drawing from prejudices against having one eye, including those brought about by physiognomy and concepts of work that equated supposedly unhealthy bodies with incapacity, advertisements such as this also buttressed such discrimination by presenting it as an unquestionable reality in public forums. Sources like this not only pandered to the demands of this rhetoric but sought to cash in on them by providing means with which to occlude physical difference from public view. For Boissonneau and his contemporaries, prejudices against physical incompleteness were what made their work viable in the competitive marketplace.

Advertisements for false teeth sold by firms such as Mr. P. B. André, Field and Co., and the Holborn Dental Institute, meanwhile, stressed the completeness of their devices as replacements for lost teeth.[114] Each of these firms guaranteed that their teeth "answer[ed] to the purpose for which they are intended, viz: Mastication, Articulation, and Natural Appearance."[115] False teeth were thus pitched as prostheses that could replace both functionally and aesthetically the body part(s) that they stood in for. As we see in Chap. 6, not everyone was convinced by such claims.

Wig makers were often much bolder in their assertions, claiming that their devices could not merely supplement but enhance. An advertisement for one seller reads as follows:

> The natural grace and adornment of these,
> Can't fail to delight, but are certain to please.
> No sooner, in fact, are they worn by the fair,
> Than at once they outrival the natural Hair.
> 'Tis surprising to notice how much they're worn!
> If a fair-one is seen at a play or a ball,
> Such "Fronts" are admired by each and by all;
> And if the good Lady should chance to be *single*,
> She'll shortly be hail'd by the *marriage-bells' jingle*![116]

This poem brings into focus the demand for full heads of hair but also shows how, unlike the other prostheses discussed in this book, artificial hair was not just used to stand in for something missing. False fronts and trusses were popular adornments for women, especially in the 1860s when the fashion for artificial hair was at its height. Hair enhancers and replacements, like false fronts, false chignons (artificial curls), wigs, toupees, and other hair pieces became so popular that, in 1849, 6200 pounds of human hair—which was the most popular manufacturing material for these bodily accoutrements—was imported to England from France alone.[117] From 1855 to 1868, sales of false hair went up by 400 per cent, showing the increasing popularity of artificiality.[118] By 1880, London alone was said to annually consume over 100,000 pounds of human hair.[119] The popularity of artificial hair at the mid-century shifted accepted standards for women's hair. The prevalence of additional hair pieces meant that, temporarily, the concept of completeness enveloped an expectation that women would have more hair than most were able to grow naturally. Although, as Ofek has demonstrated, artificial hair was a technology treated with suspicion even when it was at its most popular, its prevalence at the mid-century meant that the whole body upon which the concept of normalcy was shaped temporarily became one with an unnatural abundance of hair.[120] This trend reveals the arbitrary, constructed, and historically contingent nature of physical norms.

The way that prostheses were marketed as devices that could remove the visual presence of physical difference is of course questionable, and it is worth drawing attention to this before exploring the ways in which literature responded to these technologies. We can consider the process of prostheticizing the aberrant body to enable it to pass as a kind of rehabilitation process, a concept that irks many disability-studies scholars and activists. Not quite as abhorrent as eugenics, but based on a similar premise of erasure, rehabilitation seeks to remove physical difference from sight. As Henri-Jacques Stiker explains, "rehabilitation marks the appearance of a culture that attempts to complete the act of identification, of making identical. This act will cause the disabled to disappear, dissolve them in the greater and single social whole."[121] I do not wish to dismiss the reality of difficulties stemming from impairment or to undermine the importance of creating technologies that make the lives of people with disabilities easier or less painful. But it is an undoubted historical trend, evidenced by the marketing ploys described above, that the production of prostheses has often supported ideological—eradicating the visual presence of difference—and capitalistic rather than solely ameliorative and/or altruistic ends.

As this chapter has revealed, the nineteenth century witnessed the codification of a social system that privileged physical wholeness and marginalized those who displayed physical difference. Several historical factors encouraged and strengthened this situation, including the rise of bodily statistics, the growth of physiognomy, the development of Poor Law legislation, the solidification of ableist models of work, the advances of materialist approaches to mind and body, the publicity of bewildering symptoms of bodily loss, and the vogue for white teeth and full heads of hair. In such a society, prosthesis makers, whose businesses during this time benefited from a growing knowledge of the human body, technological developments, and innovations in hygiene and surgical procedures, cashed in on such demands for physical wholeness by providing devices that they claimed could conceal physical deficiency from public view, thereby allowing users to pass as normal. In contrast, as I will show, fictional representations of prostheses held a complex relationship with the emerging hegemony of wholeness. Chapters 4, 5, and 6 explore how this complex negotiation with the concept of wholeness played out in relation to representational factors including class, gender, and age. The next chapter, however, showcases how certain prosthesis narratives problematized the contemporary hegemony of physical completeness by imagining powerful non-normative and non-human alternatives. If we consider the use of prostheses inspired by a medical-model understanding of physical difference, the following chapters provide examples of Paul K. Longmore and Lauri Umansky's claim that "[the] 'medical model', powerful though it has been in shaping the life experiences of people with disabilities, has never gone uncontested."[122]

Notes

1. For more on *Temple Bar*, see Peter Blake, "The Paradox of a Periodical: *Temple Bar Magazine* under the Editorship of George Augustus Sala (1860–1863)," *The London Journal* 35, no. 2 (2010). For more on this short story, see Ryan Sweet, "'A Human Bundle': The Disaggregated Other at the *Fin de Siècle*," *Victorian Review* 40, no. 1 (2014).
2. Henham, "A Human Bundle," 58.
3. Lennard J. Davis, *Enforcing Normalcy: Disability, Deafness, and the Body* (New York: Verso, 1995).
4. Rosemarie Garland-Thomson, *Extraordinary Bodies: Figuring Physical Disability in American Culture and Literature* (New York: Columbia University Press, 1997), 8.
5. Sander L. Gilman, *Disease and Representation: Images of Illness from Madness to Aids* (Ithaca: Cornell University Press, 1988), 1–17.
6. Davis, *Enforcing Normalcy*; Lennard J. Davis, *Bending Over Backwards: Disability, Dismodernism, and Other Difficult Positions*. Cultural Front (New York: New York University Press, 2002); Katherine J. Kudlick, "Disability History, Power, and Rethinking the Idea of 'The Other,'" *PMLA* 120, no. 2 (2005): 557–61.
7. Davis, *Bending Over*, 50–51.
8. Anolik, *Demons of the Body*, 4.
9. Quayson, *Aesthetic Nervousness*, 20.
10. Davis, *Enforcing Normalcy*, 24.
11. Craton, *The Victorian*, 32–35; Davis, *Enforcing Normalcy*, 23–49.
12. Adolphe Quetelet, *A Treatise on Man and the Development of His Faculties*, ed. Thomas Simbert, trans. R. Knox (Edinburgh: William and Robert Chambers, 1842).
13. Craton, *The Victorian*, 32.

14. Qtd. in Theodore M. Porter, *The Rise of Statistical Thinking, 1820–1900* (Princeton: Princeton University Press, 1986), 102.
15. Craton, *The Victorian*, 33–34.
16. Davis, *Enforcing Normalcy*, 29.
17. Quetelet, *A Treatise*, 99.
18. Bruce Haley, *The Healthy Body and Victorian Culture* (Cambridge: Harvard University Press, 1978), 3.
19. Ibid., 20.
20. Kevin Berland, "Inborn Character and Free Will in the History of Physiognomy," in *Physiognomy in Profile: Lavater's Impact on European Culture*, ed. Melissa Percival and Graeme Tytler (Newark: University of Delaware Press, 2005).
21. Johann Caspar Lavater, *Essays on Physiognomy*, trans. Thomas Holcroft, 8th ed. (London: Willam Tegg & Co, 1853).
22. Sharrona Pearl, *About Faces: Physiognomy in Nineteenth-Century Britain* (Cambridge: Harvard University Press, 2010), 33.
23. Lucy Hartley, *Physiognomy and the Meaning of Expression in Nineteenth-Century Culture* (Cambridge: Cambridge University Press, 2006), 74.
24. John Graham, "Lavater's Physiognomy in England," *Journal of the History of Ideas* 22, no. 4 (1961): 562.
25. Richard Payne Knight, *An Analytical Inquiry into the Principles of Taste* (London: T. Payne & J. White, 1806), 199.
26. Zech 11:17 (Authorized Version).
27. Luke 11:34 (AV).
28. Lavater, *Essays*, 318.
29. Matt 18:9 (AV).
30. I would like to thank Sue Zemka for her help with these points.
31. Qtd. in Lavater, *Essays*, 386.
32. Ibid., 388.
33. See Chapter 1 and Jeffrey A. Brune and Daniel J. Wilson, eds. *Disability and Passing: Blurring the Lines of Identity* (Philadelphia: Temple University Press, 2013).
34. Thomas Southwood Smith, *The Philosophy of Health; or, An Exposition of the Physical and Mental Constitution of Man, with a View to the Promotion of Human Longevity and Happiness*, vol. 1 (London: Charles Knight, 1835), 1.
35. Alexander Bain, *The Senses and the Intellect*, 3rd ed. (New York: D. Appleton & Co., 1855).
36. Ibid., 10.
37. Alexander Bain, "Common Errors on the Mind," *Fortnightly Review* 4, no. 20 (1868): 160.

38. Lorraine J. Daston, "British Responses to Psycho-Physiology, 1860–1900," *Isis* 69, no. 2 (1978); Rick Rylance, *Victorian Psychology and British Culture: 1850–1880* (Oxford: Oxford University Press, 2000); Roger Smith, *Free Will and the Human Sciences in Britain, 1870–1910* (London: Pickering & Chatto, 2013).
39. William A. Cohen, *Embodied: Victorian Literature and the Senses* (Minneapolis: University of Minnesota Press, 2009), xii.
40. Henry Maudsley, *Responsibility in Mental Disease* (New York: D. Appleton, 1874), 17.
41. Ibid., 33.
42. Such a view, however, was not consistent with Maudsley. By 1895, he accepted the key role played by social habits and surroundings. See Neil Davie, "The Impact of Criminal Anthropology in Britain (1880–1918)," *Criminocorpus: Revue d'Histoire de La Justice, Des Crimes et Des Peines* 4 (2010). For more on Maudsley and the implications of his work in relation to discourses of disability, see Martha Stoddard Holmes, *Fictions of Affliction: Physical Disability in Victorian Culture* (Ann Arbor: University of Michigan Press, 2004), 66–68.
43. David G. Horn, *The Criminal Body: Lombroso and the Anatomy of Deviance* (London: Routledge, 2003), 8.
44. Ibid.
45. Ibid., 12–13, 59–86.
46. Among the most complete translations is Cesare Lombroso, *Criminal Man*, ed. and trans. Mary Gibson and Nicole Hahn Rafter (Durham: Duke University Press, 2006).
47. David T. Mitchell and Sharon L. Snyder, *Narrative Prosthesis: Disability and the Dependencies of Discourse* (Ann Arbor: University of Michigan Press, 2000); Quayson, *Aesthetic Nervousness.*
48. Davie, "The Impact"; Alfred F. Tredgold, *Mental Deficiency* (New York: William Wood, 1908).
49. Tobin Siebers, *The Mirror of Medusa*, revised ed. (Christchurch: Cybereditions, 2000), 27; Schweik, Susan M. *The Ugly Laws: Disability in Public.* New York: NYU Press, 2009.
50. Schweik, *The Ugly Laws*, 153. Also see Mary P. Ryan, *Women in Public: Between Banners and Ballots, 1825–1880* (Baltimore: Johns Hopkins University Press, 1992), 69.
51. Jan Bondeson, *A Cabinet of Medical Curiosities* (London: I.B. Tauris, 1997), 158.
52. Schweik, *The Ugly Laws.*
53. Holmes, *Fictions*, 63.
54. Ibid., 109–22.

55. Thomas Carlyle, *Past and Present* (Berkeley: University of California Press, 2005); Samuel Smiles, *Self-Help; with Illustrations of Character and Conduct* (London: S. W. Partridge & Co.: 1859).
56. John Tosh, *Manliness and Masculinities in Nineteenth-Century Britain: Essays on Gender, Family, and Empire* (Harlow: Pearson Longman, 2005), 96.
57. Erin O'Connor, *Raw Material: Producing Pathology in Victorian Culture* (Durham: Duke University Press, 2000), 104.
58. Henham, "A Human Bundle," 58.
59. Garland-Thomson, *Extraordinary Bodies*, 46.
60. Rosemarie Garland-Thomson, "The Beauty and the Freak," in *Points of Contact: Disability, Art, and Culture*, ed. Susan Crutchfield and Marcy Epstein (Ann Arbor: University of Michigan Press, 2000), 191.
61. O'Connor, *Raw Material*, 106–17.
62. William Dodd, *A Narrative of the Experience and Sufferings of William Dodd, A Factory Cripple. Written by Himself*, in *Factory Lives: Four Nineteenth-Century Working-Class Autobiographies*, ed. James R. Simmons, Jr. (London: Broadview, 2007). For more on the so-called *superfluous-woman* issue, see Nan H. Dreher, "Redundancy and Emigration: The 'Woman Question' in Mid-Victorian Britain," *Victorian Periodicals Review* 26, no. 1 (1993).
63. Wilkie Collins, *Poor Miss Finch* (Oxford: Oxford University Press, 2000); Holmes, *Fictions*, 34–73.
64. Galia Ofek, *Representations of Hair in Victorian Literature and Culture* (Aldershot: Ashgate, 2009), 2–3.
65. Qtd. in Kay Heath, *Aging by the Book: The Emergence of Midlife in Victorian Britain* (Albany: SUNY Press, 2009), 25.
66. Heath, *Aging*, 25.
67. Karen Halttunen, *Confidence Men and Painted Women: A Study of Middle-Class Culture in America, 1830–1870* (New Haven: Yale University Press, 1982), 42–43.
68. M. D. K. Bremner, *The Story of Dentistry from the Dawn of Civilization to the Present* (New York: Dental Items of Interest, 1939), 55.
69. Before this time, especially in England, dentistry was amateurish and replete with charlatanism. Its reputation was very poor, and it served primarily as a relief of pain service. N. David Richards, "Dentistry in England in the 1840s: The First Indications of a Movement Towards Professionalization," *Medical History* 12, no. 2 (1968): 139.
70. Halttunen, *Confidence Men*, 40.
71. Thomas Richards argues that out of the Great Exhibition, the 1887 and 1897 Jubilees of Queen Victoria, and the various Imperial Exhibitions of the 1890s "came all of the familiar imperatives of modern commodity

culture, with its emphasis on status, exoticism, health, and female sexuality." Richards plots the rise of advertising against these public spectacles while also considering the importance of illustrations, which were printed more cheaply following developments in printing technologies. Thomas Richards, *The Commodity Culture of Victorian England: Advertising and Spectacle, 1851–1914* (Stanford: Stanford University Press, 1990), 71.

72. M. Cox et al., "Dental History: Eighteenth and Nineteenth Century Dental Restoration, Treatment and Consequences in a British Nobleman," *British Dental Journal* 189, no. 11 (2000): 595.
73. Peter A. Reichart, "Toothpastes Containing Betel Nut (Areca Catechu L.) from England of the Nineteenth Century," *Journal of the History of Medicine* 39 (1984).
74. "The Teeth," *The London Reader: Of Literature, Science, Art and General Information* 29, no. 743 (1877): 305.
75. Silas Weir Mitchell, "Phantom Limbs," *Lippincott's Monthly Magazine* 8 (1871): 565.
76. Ibid., 565.
77. Ibid., 567–68.
78. Ibid., 567.
79. Ibid., 564.
80. Stanley Finger and Meredith Hustwit, "Five Early Accounts of Phantom Limb in Context: Paré, Descartes, Lemos, Bell, and Mitchell," *Neurosurgery* 52, no. 3 (2003): 675–86.
81. Silas Weir Mitchell, "The Case of George Dedlow," in *The Autobiography of a Quack and the Case of George Dedlow*, by Silas Weir Mitchell (New York: Century, 1900).
82. Mitchell, "Phantom Limbs," 564.
83. Mitchell, "The Case," 131.
84. Ibid., 132; Mitchell, "Phantom Limbs," 566.
85. O'Connor, *Raw Material*, 106–17.
86. Silas Weir Mitchell, *Injuries of the Nerves and Their Consequences* (New York: Dover, 1965).
87. O'Connor, *Raw Material*, 106–17.
88. John Erichsen, *The Science and Art of Surgery: Being a Treatise on Surgical Injuries, Diseases, and Operations*, 2nd ed. (London: Walton and Maberly, 1857), 87; qtd. in O'Connor *Raw Material*, 107–108.
89. O'Connor *Raw Material*, 107–108.
90. For more on this condition, see G. Martin Huggins, *Amputation Stumps: Their Care and After-Treatment* (London: Henry Frowde, Hodder and Stoghton, 1918), 149; O'Connor, *Raw Material*, 109; and Julie Anderson, "'Jumpy Stump': Amputation and Trauma in the First World War," *First World War Studies* 6, no. 1 (2015).

91. Jane Wood, *Passion and Pathology in Victorian Fiction* (Oxford: Oxford University Press, 2001), 12.
92. Frederick Marryat, *Jacob Faithful* (London: Richard Edward King, 1834); Charles Dickens, *Our Mutual Friend* (London: Everyman, 2000).
93. Charles Dickens, *Dombey and Son* (London: Penguin, 2002).
94. Marryat *Jacob Faithful*, 178.
95. For more on this fascination, see Rosemarie Garland-Thomson, *Freakery: Cultural Spectacles of the Extraordinary Body* (New York: New York University Press, 1996); Garland-Thomson, *Extraordinary Bodies*; Rachel Adams, *Sideshow U.S.A.: Freaks and the American Cultural Imagination* (Chicago: University of Chicago Press, 2001); Craton, *The Victorian*; Marlene Tromp, ed., *Victorian Freaks: The Social Context of Freakery in Britain* (Columbus: Ohio State University Press, 2007); and Nadja Durbach, *Spectacle of Deformity: Freak Shows and Modern British Culture* (Berkeley: University of California Press, 2009).
96. Gilman, *Disease and Representation.*
97. Michael Cotsell, *The Companion to Our Mutual Friend* (London: Allen and Unwin, 1986), 50.
98. Dickens, *Our Mutual Friend*, 513–15.
99. Herbert Sussman and Gerhard Joseph, "Prefiguring the Posthuman: Dickens and Prosthesis," *Victorian Literature and Culture* 32, no. 2 (2004).
100. See Kara Rogers, "9 Dickensian Diagnoses," *Encyclopedia Britannica*, accessed June 15, 2018, https://www.britannica.com/list/9-dickensian-diagnoses; Lisa Sanders, "Diagnosing with Dickens," *The New York Times*, December 17, 2006, https://www.nytimes.com/2006/12/17/magazine/17wwln_diagnosis.t.html; Meir Kryger, "Charles Dickens: Impact on Medicine and Society," *Journal of Clinical Sleep Medicine* 8, no. 3 (2012); and Hannah Devlin, "Charles Dickens' Contribution to Medicine Highlighted in London Exhibition," *The Guardian*, May 16, 2018, http://www.theguardian.com/science/2018/may/16/dickens-acute-observations-on-health-celebrated-in-london-exhibition.
101. Dickens, *Our Mutual Friend*, 151.
102. Mitchell, "Phantom Limbs," 564.
103. Dickens, *Our Mutual Friend*, 515.
104. Here I suggest that Dickens feminizes Wegg in this scene, but, like Talia Schaffer, I do not endorse the tired and problematic reading of disability as a symbol for castration. For a through critique of this familiar reading, see Talia Schaffer, *Romance's Rival: Familiar Marriage in Victorian Fiction* (Oxford: Oxford University Press, 2016), 159–98.
105. Charles Dickens, *Nicholas Nickleby*, 2 vols. (London: Scholar Press, 1982).

106. Philip Hobsbaum, *A Reader's Guide to Charles Dickens* (Syracuse: Syracuse University Press, 1998), 52.
107. For more on the complexity and inconsistency of Dickens's treatment of disability, see Stanley F. Wainapel, "Dickens and Disability," *Disability and Rehabilitation* 18, no. 12 (1996); Adrienne E. Gavin, "Dickens, Wegg, and Wooden Legs," *Our Mutual Friend: The Scholarly Pages*, accessed June 18, 2018, http://omf.ucsc.edu/london-1865/victorian-city/wooden-legs.html; Elisabeth G. Gitter, "The Blind Daughter in Charles Dickens's Cricket on the Hearth," *SEL Studies in English Literature 1500–1900* 39, no. 4 (1999); Holmes, *Fictions*; Sussman and Joseph, "Prefiguring the Posthuman"; Julia Miele Rodas, "Tiny Tim, Blind Bertha, and the Resistance of Miss Mowcher: Charles Dickens and the Uses of Disability," *Dickens Studies Annual* 34 (2004); Tom Shakespeare, "Stickin' up for Dickens," *Ouch! It's a Disability Thing*, last modified December 13, 2006, http://www.bbc.co.uk/ouch/opinion/stickin_up_for_dickens.shtml; Paul Marchbanks, "From Caricature to Character: The Intellectually Disabled in Dickens's Novels," *Dickens Quarterly* 23 (2006); Christine Ferguson, "Sensational Dependence: Prosthesis and Affect in Dickens and Braddon," *Lit: Literature Interpretation Theory* 19, no. 1 (2008); Karen Bourrier, "Reading Laura Bridgman: Literacy and Disability in Dickens's *American Notes*," *Dickens Studies Annual: Essays on Victorian Fiction* 40, no. 1 (2009); Craton, *The Victorian*; Jennifer Esmail, "'I listened with my eyes': Writing Speech and Reading Deafness in the Fiction of Charles Dickens and Wilkie Collins," *ELH* 78, no. 4 (2011); and Heather Tilley, "The Sentimental Touch: Dickens's *Old Curiosity Shop* and the Feeling Reader," *Journal of Victorian Culture* 16, no. 2 (2011).
108. Thomas, *The Commodity*; Claire L. Jones, *The Medical Trade Catalogue, 1870–1914* (London: Pickering & Chatto, 2014); James F. Stark, "'Recharge My Exhausted Batteries': Overbeck's Rejuvenator, Patenting, and Public Medical Consumers, 1924–37," *Medical History* 58, 4 (2014). Also see James Harvey Young, *The Toadstool Millionaires: A Social History of Patent Medicines in America before Federal Regulation* (Princeton: Princeton University Press, 1961); and Anne Digby, *Making a Medical Living: Doctors and Patients in the English Market for Medicine, 1720–1911* (Cambridge: Cambridge University Press, 2002).
109. O'Connor, *Raw Material*, 102–47.
110. Qtd. in George E. Marks, *Marks' Patent Artificial Limbs with Rubber Hands and Feet* (New York: A. A. Marks, 1888), 154.
111. Qtd. in ibid., 232.
112. Jones, *The Medical*, 38.

113. Auguste Boissonneau, Advertisement for Auguste Boissonneau's Artificial Eyes, *Freeman's Journal and Daily Commercial Advertiser*, August 26, 1852, front page.
114. P. B. André, "Advertisement for Mr. P. B. André, 'Surgeon & Mechanical Dentist,'" n.d., *British Library Online Gallery*, accessed June 18, 2018, http://www.bl.uk/onlinegallery/onlineex/evancoll/a/014eva000000000u06984000.html; Field & Co., Advertisement for Artificial Teeth Patented by Field & Co., 1884, *British Library Online Gallery*, accessed June 18, 2018, http://www.bl.uk/onlinegallery/onlineex/evancoll/a/zoomify72552.html; Holborn Dental Institute, Advertisement for the Holborn Dental Institute, 1886, *British Library Online Gallery*, accessed June 18, 2018, http://www.bl.uk/onlinegallery/onlineex/evancoll/a/014eva000000000u05139000.html.
115. Holborn Dental Institute, Advertisement.
116. F. Browne, Advertisement for Professor F. Browne's Ladies' Toupees, n.d., EPH 154, Hair care box 1, Wellcome Lib., London.
117. Alexander Rowland, *The Human Hair, Popularly and Physiologically Considered with Special Reference to Its Preservation, Improvement and Adornment, and the Various Modes of Its Decoration in All Countries* (London: Piper, Brothers & Co., 1853), 164.
118. Ofek, *Representations*, 37.
119. C. Henri Leonard, *The Hair: Its Growth, Care, Diseases and Treatment* (Detroit: C. Henri Leonard, 1880), 9.
120. Ofek, *Representations.*
121. Henri-Jacques Stiker, *A History of Disability*, trans. William Sayers, revised edition (Ann Arbor: University of Michigan Press, 1999), 128.
122. Paul K. Longmore and Lauri Umansky, *The New Disability History: American Perspectives* (New York: New York University Press, 2001), 259.

References

Adams, Rachel. *Sideshow U.S.A.: Freaks and the American Cultural Imagination.* Chicago: University of Chicago Press, 2001.

André, P. B. "Advertisement for Mr. P. B. André, 'Surgeon & Mechanical Dentist.'" n.d. *British Library Online Gallery.* Accessed June 18, 2018. http://www.bl.uk/onlinegallery/onlineex/evancoll/a/014eva000000000u06984000.html.

Anderson, Julie. "'Jumpy Stump': Amputation and Trauma in the First World War." *First World War Studies* 6, no. 1 (2015): 9–19.

Anolik, Ruth Bienstock, ed. *Demons of the Body and Mind: Essays on Disability in Gothic Literature.* Jefferson: McFarland & Company, 2010.

Bain, Alexander. *The Senses and the Intellect.* 3rd ed. New York: D. Appleton & Co., 1855.

———. "Common Errors on the Mind." *Fortnightly Review* 4, no. 20 (1868): 160–75.

Berland, Kevin. "Inborn Character and Free Will in the History of Physiognomy." In *Physiognomy in Profile: Lavater's Impact on European Culture*, edited by Melissa Percival and Graeme Tytler, 25–38. Newark: University of Delaware Press, 2005.

Blake, Peter. "The Paradox of a Periodical: *Temple Bar* Magazine under the Editorship of George Augustus Sala (1860–1863)." *The London Journal* 35, no. 2 (2010): 185–209.

Boissonneau, Auguste. Advertisement for Auguste Boissonneau's Artificial Eyes. *Freeman's Journal and Daily Commercial Advertiser*, 26 August 1852.

Bondeson, Jan. *A Cabinet of Medical Curiosities.* London: I.B. Tauris, 1997.

Bourrier, Karen. "Reading Laura Bridgman: Literacy and Disability in Dickens's *American Notes.*" *Dickens Studies Annual: Essays on Victorian Fiction* 40, no. 1 (2009): 37–60.

Bremner, M. D. K. *The Story of Dentistry from the Dawn of Civilization to the Present.* New York: Dental Items of Interest, 1939.

Browne, F. Advertisement for Professor F. Browne's Ladies' Toupees. n.d. EPH 154. Hair care box 1. Wellcome Lib., London.

Brune, Jeffrey A., and Daniel J. Wilson, eds. *Disability and Passing: Blurring the Lines of Identity.* Philadelphia: Temple University Press, 2013.

Carlyle, Thomas. *Past and Present.* Berkeley: University of California Press, 2005.

Cohen, William A. *Embodied: Victorian Literature and the Senses.* Minneapolis: University of Minnesota Press, 2009.

Collins, Wilkie. *Poor Miss Finch.* Oxford: Oxford University Press, 2000.

Cotsell, Michael. *The Companion to Our Mutual Friend.* London: Allen and Unwin, 1986.

Cox, M., J. Chandler, A. Boyle, P. Kneller, and R. Haslam. "Dental History: Eighteenth and Nineteenth Century Dental Restoration, Treatment and Consequences in a British Nobleman." *British Dental Journal* 189, no. 11 (2000): 593–96.

Craton, Lillian. *The Victorian Freak Show: The Significance of Disability and Physical Differences in 19th-Century Fiction.* New York: Cambria Press, 2009.

Daston, Lorraine J. "British Responses to Psycho-Physiology, 1860–1900." *Isis* 69, no. 2 (1978): 192–208.

Davie, Neil. "The Impact of Criminal Anthropology in Britain (1880–1918)." *Criminocorpus: Revue d'Histoire de La Justice, Des Crimes et Des Peines* 4 (2010): n.pag.

Davis, Lennard J. *Enforcing Normalcy: Disability, Deafness, and the Body.* New York: Verso, 1995.

———. *Bending Over Backwards: Disability, Dismodernism, and Other Difficult Positions*. Cultural Front. New York: New York University Press, 2002.

Devlin, Hannah. "Charles Dickens' Contribution to Medicine Highlighted in London Exhibition." *The Guardian*, May 16, 2018. http://www.theguardian.com/science/2018/may/16/dickens-acute-observations-on-health-celebrated-in-london-exhibition.

Dickens, Charles. *Nicholas Nickleby*. 2 vols. London: Scholar Press, 1982.

———. *Dombey and Son*. London: Penguin, 2002.

———. *Our Mutual Friend*. London: Everyman, 2000.

Digby, Anne. *Making a Medical Living: Doctors and Patients in the English Market for Medicine, 1720–1911*. Cambridge: Cambridge University Press, 2002.

Dodd, William. *A Narrative of the Experience and Sufferings of William Dodd, A Factory Cripple. Written by Himself.* In *Factory Lives: Four Nineteenth-Century Working-Class Autobiographies*, edited by James R. Simmons, Jr., 181–300. London: Broadview, 2007.

Dreher, Nan H. "Redundancy and Emigration: The 'Woman Question' in Mid-Victorian Britain." *Victorian Periodicals Review* 26, no. 1 (1993): 3–7.

Durbach, Nadja. *Spectacle of Deformity: Freak Shows and Modern British Culture*. Berkeley: University of California Press, 2009.

Erichsen, John. *The Science and Art of Surgery: Being a Treatise on Surgical Injuries, Diseases, and Operations*. 2nd ed. London: Walton and Maberly, 1857.

Esmail, Jennifer. "'I listened with my eyes': Writing Speech and Reading Deafness in the Fiction of Charles Dickens and Wilkie Collins." *ELH* 78, no. 4 (2011): 991–1020.

Ferguson, Christine. "Sensational Dependence: Prosthesis and Affect in Dickens and Braddon." *Lit: Literature Interpretation Theory* 19, no. 1 (2008): 1–25.

Field & Co. Advertisement for Artificial Teeth Patented by Field & Co. 1884. British Library Online Gallery. Accessed June 18, 2018. http://www.bl.uk/onlinegallery/onlineex/evancoll/a/zoomify72552.html.

Finger, Stanley, and Meredith Hustwit. "Five Early Accounts of Phantom Limb in Context: Paré, Descartes, Lemos, Bell, and Mitchell." *Neurosurgery* 52, no. 3 (2003): 675–86.

Garland-Thomson, Rosemarie. *Extraordinary Bodies: Figuring Physical Disability in American Culture and Literature*. New York: Columbia University Press, 1997.

———. *Freakery: Cultural Spectacles of the Extraordinary Body*. New York: New York University Press, 1996.

———. "The Beauty and the Freak." In *Points of Contact: Disability, Art, and Culture*, edited by Susan Crutchfield and Marcy Epstein, 181–96. Ann Arbor: University of Michigan Press, 2000.

Gavin, Adrienne E. "Dickens, Wegg, and Wooden Legs." *Our Mutual Friend*: The Scholarly Pages. Accessed June 18, 2018. http://omf.ucsc.edu/london-1865/victorian-city/wooden-legs.html.

Gilman, Sander L. *Disease and Representation: Images of Illness from Madness to Aids*. Ithaca: Cornell University Press, 1988.

Gitter, Elisabeth G. "The Blind Daughter in Charles Dickens's Cricket on the Hearth." *SEL Studies in English Literature 1500–1900* 39, no. 4 (1999): 675–89.

Graham, John. "Lavater's Physiognomy in England." *Journal of the History of Ideas* 22, no. 4 (1961): 561–72.

Haley, Bruce. *The Healthy Body and Victorian Culture*. Cambridge: Harvard University Press, 1978.

Halttunen, Karen. *Confidence Men and Painted Women: A Study of Middle-Class Culture in America, 1830–1870*. New Haven: Yale University Press, 1982.

Hartley, Lucy. *Physiognomy and the Meaning of Expression in Nineteenth-Century Culture*. Cambridge: Cambridge University Press, 2006.

Heath, Kay. *Aging by the Book: The Emergence of Midlife in Victorian Britain*. Albany: SUNY Press, 2009.

Henham, Ernest G. "A Human Bundle." *Temple Bar* 111, no. 438 (1897): 42–58.

Hobsbaum, Philip. *A Reader's Guide to Charles Dickens*. Syracuse: Syracuse University Press, 1998.

Holborn Dental Institute. Advertisement for the Holborn Dental Institute. 1886. British Library Online Gallery. Accessed June 18, 2018. http://www.bl.uk/onlinegallery/onlineex/evancoll/a/014eva000000000u05139000.html.

Holmes, Martha Stoddard. *Fictions of Affliction: Physical Disability in Victorian Culture*. Ann Arbor: University of Michigan Press, 2004.

Hood, Thomas. *Miss Kilmansegg and Her Precious Leg*. In *Selected Poems of Thomas Hood*, edited by John Clubbe, 193–270. Cambridge: Harvard University Press, 1970.

Horn, David G. *The Criminal Body: Lombroso and the Anatomy of Deviance*. London: Routledge, 2003.

Huggins, G. Martin. *Amputation Stumps: Their Care and After-Treatment*. London: Henry Frowde, Hodder and Stoghton, 1918.

Jones, Claire L. *The Medical Trade Catalogue, 1870–1914*. London: Pickering & Chatto, 2014.

Knight, Richard Payne. *An Analytical Inquiry into the Principles of Taste*. London: T. Payne & J. White, 1806.

Kryger, Meir. "Charles Dickens: Impact on Medicine and Society." *Journal of Clinical Sleep Medicine* 8, no. 3 (2012): 333–38.

Kudlick, Katherine J. "Disability History, Power, and Rethinking the Idea of 'The Other.'" *PMLA* 120, no. 2 (2005): 557–61.

Lavater, Johann Caspar. *Essays on Physiognomy*. Translated by Thomas Holcroft. 8th ed. London: Willam Tegg & Co, 1853.

Leonard, C. Henri. *The Hair: Its Growth, Care, Diseases and Treatment*. Detroit: C. Henri Leonard, 1880.

Lewis, Henry Clay. "The Indefatigable Bear Hunter." In *Louisiana Swamp Doctor: The Writings of Henry Clay Lewis*, edited by John Q. Anderson, 233–44. Baton Rouge: Louisiana State University Press, 1962.

Lombroso, Cesare. *Criminal Man*. Edited and translated by Mary Gibson and Nicole Hahn Rafter. Duke University Press, 2006.

Longmore, Paul K., and Lauri Umansky. *The New Disability History: American Perspectives*. New York: New York University Press, 2001.

Marchbanks, Paul. "From Caricature to Character: The Intellectually Disabled in Dickens's Novels." *Dickens Quarterly* 23 (2006): 3–14; 67–85; 169–81.

Marks, George E. *Marks' Patent Artificial Limbs with Rubber Hands and Feet*. New York: A. A. Marks, 1888.

Marryat, Frederick. *Jacob Faithful*. London: Richard Edward King, 1834.

Maudsley, Henry. *Responsibility in Mental Disease*. New York: D. Appleton, 1874.

Mitchell, David T., and Sharon L. Snyder. *Narrative Prosthesis: Disability and the Dependencies of Discourse*. Ann Arbor: University of Michigan Press, 2000.

Mitchell, Silas Weir. "The Case of George Dedlow." In *The Autobiography of a Quack and the Case of George Dedlow*, by Silas Weir Mitchell, 113–49. New York: Century, 1900.

———. "Phantom Limbs." *Lippincott's Monthly Magazine* 8 (1871): 563–69.

———. *Injuries of the Nerves and Their Consequences*. New York: Dover, 1965.

O'Connor, Erin. *Raw Material: Producing Pathology in Victorian Culture*. Durham: Duke University Press, 2000.

Ofek, Galia. *Representations of Hair in Victorian Literature and Culture*. Aldershot: Ashgate, 2009.

Pearl, Sharrona. *About Faces: Physiognomy in Nineteenth-Century Britain*. Cambridge: Harvard University Press, 2010.

Porter, Theodore M. *The Rise of Statistical Thinking, 1820–1900*. Princeton: Princeton University Press, 1986.

Quayson, Ato. *Aesthetic Nervousness: Disability and the Crisis of Representation*. New York: Columbia University Press, 2007.

Quetelet, Adolphe. *A Treatise on Man and the Development of His Faculties*. Edited by Thomas Simbert. Translated by R. Knox. Edinburgh: William and Robert Chambers, 1842.

Reichart, Peter A. "Toothpastes Containing Betel Nut (Areca Catechu L.) from England of the Nineteenth Century." *Journal of the History of Medicine* 39 (1984): 65–68.

Richards, N. David. "Dentistry in England in the 1840s: The First Indications of a Movement Towards Professionalization." *Medical History* 12, no. 2 (1968): 137–52.

Richards, Thomas. *The Commodity Culture of Victorian England: Advertising and Spectacle, 1851–1914*. Stanford: Stanford University Press, 1990.

Rodas, Julia Miele. "Tiny Tim, Blind Bertha, and the Resistance of Miss Mowcher: Charles Dickens and the Uses of Disability." *Dickens Studies Annual* 34 (2004): 51–94.

Rogers, Kara. "9 Dickensian Diagnoses." *Encyclopedia Britannica*. Accessed June 15, 2018. https://www.britannica.com/list/9-dickensian-diagnoses.

Rowland, Alexander. *The Human Hair, Popularly and Physiologically Considered with Special Reference to Its Preservation, Improvement and Adornment, and the Various Modes of Its Decoration in All Countries*. London: Piper, Brothers & Co., 1853.

Ryan, Mary P. *Women in Public: Between Banners and Ballots, 1825–1880*. Baltimore: Johns Hopkins University Press, 1992.

Rylance, Rick. *Victorian Psychology and British Culture: 1850–1880*. Oxford: Oxford University Press, 2000.

Sanders, Lisa. "Diagnosing with Dickens." *The New York Times*, December 17, 2006. https://www.nytimes.com/2006/12/17/magazine/17wwln_diagnosis.t.html.

Schaffer, Talia. *Romance's Rival: Familiar Marriage in Victorian Fiction*. Oxford: Oxford University Press, 2016.

Schweik, Susan M. *The Ugly Laws: Disability in Public*. New York: NYU Press, 2009.

Shakespeare, Tom. "Stickin' up for Dickens." *Ouch! It's a Disability Thing*. Last modified December 13, 2006. http://www.bbc.co.uk/ouch/opinion/stickin_up_for_dickens.shtml.

Siebers, Tobin. *The Mirror of Medusa*. Revised ed. Christchurch: Cybereditions, 2000.

Smiles, Samuel. *Self-Help; with Illustrations of Character and Conduct*. London: S. W. Partridge & Co.: 1859.

Smith, Roger. *Free Will and the Human Sciences in Britain, 1870–1910*. London: Pickering & Chatto, 2013.

Smith, Thomas Southwood. *The Philosophy of Health; or, An Exposition of the Physical and Mental Constitution of Man, with a View to the Promotion of Human Longevity and Happiness*. Vol. 1. London: Charles Knight, 1835.

Stark, James F. "'Recharge My Exhausted Batteries': Overbeck's Rejuvenator, Patenting, and Public Medical Consumers, 1924–37." *Medical History* 58, 4 (2014): 498–518.

Stiker, Henri-Jacques. *A History of Disability*. Translated by William Sayers. Revised edition. Ann Arbor: University of Michigan Press, 1999.

Sussman, Herbert, and Gerhard Joseph. "Prefiguring the Posthuman: Dickens and Prosthesis." *Victorian Literature and Culture* 32, no. 2 (2004): 617–28.

Sweet, Ryan. "'A Human Bundle': The Disaggregated Other at the *Fin de Siècle*." *Victorian Review* 40, no. 1 (2014): 14–18.

"The Teeth". *The London Reader: Of Literature, Science, Art and General Information* 29, no. 743 (1877): 305.

Tilley, Heather. "The Sentimental Touch: Dickens's *Old Curiosity Shop* and the Feeling Reader." *Journal of Victorian Culture* 16, no. 2 (2011): 226–41.
Tosh, John. *Manliness and Masculinities in Nineteenth-Century Britain: Essays on Gender, Family, and Empire*. Harlow: Pearson Longman, 2005.
Tredgold, Alfred F. *Mental Deficiency*. New York: William Wood, 1908.
Tromp, Marlene, ed. *Victorian Freaks: The Social Context of Freakery in Britain*. Columbus: Ohio State University Press, 2007.
Wainapel, Stanley F. "Dickens and Disability." *Disability and Rehabilitation* 18, no. 12 (1996): 629–32.
Wood, Jane. *Passion and Pathology in Victorian Fiction*. Oxford: Oxford University Press, 2001.
Young, James Harvey. *The Toadstool Millionaires: A Social History of Patent Medicines in America before Federal Regulation*. Princeton: Princeton University Press, 1961.

CHAPTER 3

"The Infurnal Thing": Autonomy and Ability in Narratives of Disabling, Self-acting, and Weaponized Prostheses

Though not a prosthesis in quite the same way as an artificial leg, the mechanical trousers of Nick Park's 1993 Academy Award–winning animated short *Wallace and Gromit: The Wrong Trousers* provides a useful entry point for thinking about themes of autonomy and ability in relation to replacement body parts.[1] In Park's imaginative universe, Wallace purchases a pair of ex-NASA mechanical trousers for his companion dog Gromit as a birthday present. Not quite a direct physical replacement for Wallace's legs but certainly a functional supplement, unfortunately for Park's eponymous characters, these trousers do not behave as they should. Riffing off the popular sci-fi trope of the machine-gone-bad— *à la* the *Terminator* franchise—the controls for the trousers end up in the wrong hands (or wings) as the device is used to perform a bank robbery by the criminal mastermind penguin, Feathers McGraw. Most distressingly for Wallace, this feat is performed while he is asleep wearing the so-called techno-trousers. This story is clearly more fun than serious, but what cannot be denied is that part of the representational force of the piece stems from the way that it playfully engages with genuine human anxieties, namely those related to a body part holding the body whole to ransom—with technology usurping the human as master, and with technological improvements rendering some more able than others. As I argue here, these very themes were hallmarks of representations of self-acting and weaponized prostheses over a century before Park's popular animation.

R. Sweet, *Prosthetic Body Parts in Nineteenth-Century Literature and Culture*, Palgrave Studies in Nineteenth-Century Writing and Culture, https://doi.org/10.1007/978-3-030-78589-5_3

The focus of this chapter is on autonomy as a framing concept in nineteenth- and early twentieth-century representations of prosthetic body parts. By placing imaginaries of devices such as artificial legs, arms, eyes, and teeth in context with contextual factors that brought understandings of human and machine agency under the spotlight, I maintain that several prosthesis narratives problematized the contemporary hegemony of physical wholeness (explored in the previous chapter) by imaging powerful non-normative and non-human alternatives—human-machine splices that are enhanced by prosthetic technologies. These parts challenge the autonomy of the host, the body rendered whole by prosthesis. For this reason, my discussion unveils a paradox: stories that focus upon prosthetics—devices underpinned by an ableist medical-model understanding of bodily difference, which are used to normalize deviant bodies—often challenge the very philosophy that has brought about the use of prosthetic body parts. These tales mobilize bodily alternatives that compete with hegemonic norms. The following analysis both develops and contrasts Erin O'Connor's argument that "at the body-machine interface lies a 'prosthetic territory,' a frontier of potential resistance whose liberatory effects derive, paradoxically, from a strategic complicity with and dependence on machines."[2] I show how prosthetics were imagined as both liberating and restricting, and how, in many cases, the human-machine splice that the use of these devices engendered was imagined as a threat to the organic body. In each case, the autonomy of prosthesis-using subjects (seeming cyborgs in certain instances) or the powerful prosthetic device that they use, or are used by, challenged the preference for physical wholeness by imagining the possible results that such social pressures could bring about: life forms more potent than normal human subjects.

In terms of its engagement with prostheses, devices sometimes imagined as kinds of machines in the literature of the nineteenth century, my analysis provides further evidence for Tamara Ketabgian's observations regarding the animacy of machines in the Victorian imagination:

> [M]achines were not simply soulless, lifeless, predictable, and unidimensional; not simply opposed to organic feeling and vitality; and not simply reductive material objects—if objects are ever so. They lead such a rich figurative life, yielding a broad literary array of habits, feelings, communities, and subjectivities. As science and technology studies have shown, these engines served as coordinated dynamic networks, with systems of complex interdependence that formatively shaped physiological and thermodynamic models of life.[3]

I explore the perceived autonomy of the mechanical prosthetic body part, particularly in a section where I analyse the self-acting prostheses that appear in Frances Parker, Countess of Morley's *The Flying Burgermaster* (1832), the popular song "Cork Leg" (published by John Ashton in 1888 but in circulation for many years before), and J. Stuart Blackton's short film *The Thieving Hand* (1908).[4] In these instances, the representation of the prosthetic body part foreshadows Jacques Derrida's warning about the "supplement" (for him, writing), a signifier that he calls a dangerous since it "claims to be presence and the sign of the thing itself." As Derrida contends: "It is the addition of a technique, a sort of artificial and artful ruse to make speech present when it is actually absent."[5] The supplement is dangerous as it "threatens to subvert and supplant the whole by questioning the whole's structural integrity."[6] This same dynamic can be seen in nineteenth-century prosthesis narratives. In the literary representations in question, users are often either rendered less human by using primitive forms of prosthesis or made slaves to the will of hyper-sophisticated devices.

Human-Machine Minds and Bodies

The anxieties relating to prosthesis and autonomy conflate various nineteenth-century cultural misgivings surrounding human-machine and mind-body relationships, machine autonomy, and ontology. As numerous critics have observed, a significant amount of nineteenth-century cultural attention to these topics revealed alarm about a loss of human agency in light of the materialist "disenchantment" of life and the unrelenting forward march of technology.[7] The emergence of self-acting machines, such as the Jacquard loom and Charles Babbage's Difference Engine, encouraged contemporaries to reconsider the subject/object binary. The use of newly fashioned self-feedback loops to produce automatic processes made machines appear more lifelike, while, thanks to a host of influential evolutionary and materialist theories, humans increasingly came to be understood as almost machine-like—"conscious automata" as psycho-physiologist Thomas Henry Huxley would argue in 1874.[8]

Many were optimistic about the rise of empirical science and about the productive potential of new, self-acting technologies. The Scottish professor Andrew Ure, a figure that many consider to be the chief apologist of nineteenth-century machinery, was a notable example. In *The Philosophy of Manufacture* (1835), he raved about a self-acting mule (a machine invented in the late eighteenth century to effect more efficient yarn spinning), which he claimed was "a creation destined to restore order among the industrious classes, and to confirm Great Britain in the empire of art."[9] Babbage was similarly sanguine about the utilitarian benefits of machinery. In *An Essay on the General Principles Which Regulate the Application of Machinery to Manufactures and the Mechanical Arts* (1827), he talked highly of the reparative capacity of man's fusion with technology: "a less general use of tools for human hands, is to assist the labour of those who are deprived by nature, or by accident, of some of their limbs. … These triumphs of skill and ingenuity deserve a double portion of our admiration when applied to mitigate the severity of natural or accidental misfortune."[10] For technophiles such as Babbage, industrial machines offered an alternative solution to the crisis of physical loss. By mobilizing a framework of human-machine splicing, Babbage's techno-optimistic view was linked to but different from the popular view that concealing prostheses provided cures for perceived physical losses. Concealing physical differences to look able would become less necessary if industrial machines could be operated by non-normative bodies.

In addition to the voices of technophiles, such as Ure and Babbage, who encouraged the rise of intelligent machines, theories also emerged that sought to redefine the human body in mechanical terms. The growth of the new science of thermodynamics, the branch of physics that dealt with the relationship between heat and other types of energy, led to a radical reconceptualization of the active human as fundamentally an engine. Jenny Uglow suggests that these theories harked back to Julien Offray de La Mettrie's *L'Homme Machine* (1747), which (partly provoked by René Descartes's definition of animals as "automata") attempted to "restate the problem of the mind as physical, not meta-physical—to define man as a mechanical entity in which perception, emotion, understanding, foresight and thought were produced by organic causes."[11] In a similar way to how Ure and Babbage talked about intelligent machines, as Anson Rabinbach explains, scientists also linked the understanding of humans as machines to a drive for progress: "If the working body was a motor, some scientists reasoned, it might even be possible to eliminate the stubborn resistance to perpetual work that distinguished the human body from a machine."[12] Such a technological forward march was also promised by contemporary psycho-physiologists, the forerunners of modern psychology, who from the 1870s onwards encouraged a reconsideration of the mind-body relationship. They suggested that, like the body, the mind is subject to causes and actions instigated by changes in force and matter. As several critics have observed, this materialist philosophy proved controversial because it questioned the existence of both human free will and divine agency, suggesting instead that human action is determined by evolutionary, environmental, and physical causes.[13] While proponents of psycho-physiology, such as Herbert Spencer, Alexander Bain, Henry Maudsley, William Kingdon Clifford, and Huxley, recognized the possible ethical and religious implications of undermining the existence of human free will (they often posed dualistic theories that sought to encourage a materialist understanding of the mind while simultaneously encouraging readers to respond ethically to the determining forces of nature), many remained resistant to this philosophical position, which seemed to render humans equivalent to machines despite each theorist's careful treatment of the concept of free will.

The most extreme resistance to "the rise of the machine" occurred earlier in the century, during what has been named the Luddite movement. While the actions of the Luddites responded less to the philosophical dehumanizing of humanity than it did to the physical implications and social injustices caused by technological innovation, agency was nonetheless a major concern to these revolutionaries. To these radicals, technology was robbing human workers of their freedom since machines were beginning to take the place of people in the industrial workplace. As a response to the perceived crisis of unemployment effected by the rise of self-regulating machines, such as the Jacquard loom, working-class men—who claimed to be led by the mythical King Ludd—broke into factories and destroyed machinery. At least one factory owner was killed in these violent protests. This technological resistance lasted roughly from 1812 to 1817, but the legacy of technophobia endured for many years afterwards. Even today, technophobes are sometimes described as "Luddites."

After the Luddite movement subsided in 1817, a cultural resistance to the rise of science and technology remained prominent. One year after the Luddite movement ended, Mary Shelley published *Frankenstein, or the Modern Prometheus* (1818), a story that is widely held to deliver "a powerful anti-science diatribe that still reverberates as a quintessential parable of the dangers unleashed by technological creation and irresponsible scientists."[14] In later cultural representations of human interaction with new technologies, including the steam engine, a more complex relationship was often displayed. As Nicholas Daly reminds us, a great number of plays premiered in the 1860s, for example, Dion Boucicault's *After Dark*

(1868), which relied on a "cultural imaginary in which the impact of the machine, or industrial modernity more generally, on the human is a source of trepidation, or even terror, though also of fascination."[15] Along similar lines, Charles Dickens was another prominent cultural figure who displayed an acute awareness of the potential negative effects of new technology. He was "exhilarated by speed and novelty" of the railway, but "remain[ed] sceptical about the social consequences of applied technology on this scale."[16] Dickens's mixed feelings towards the railway are exemplified in both *Dombey and Son* (1846–1848) and the famous ghost story "The Signal Man" (1866).[17] Cultural apprehensions about the forward march of technology were again brought to the attention of the reading public by a series of science-fiction novels written by H. G. Wells in the 1890s, including *The Time Machine* (1895), *The Island of Dr. Moreau* (1896), *The Invisible Man* (1897), and *When the Sleeper Wakes* (1899).[18] Each of these novels depicts the potentially dystopic results of scientific experimentation or technological development gone wrong. The nervousness in many of these cultural portrayals of technology centres on a fear relating to diminishing human agency. The worry for many was that the increasing intimacy between human and machine, which saw a rebalancing of the user-technology hierarchy, could lead to the technological part usurping the human user, rendering him or her either redundant or mechanical. Enmeshed in debates surrounding the rise of technology and an increasing understanding of the human in materialist terms, what follows explores fictional responses to the physical splicing of human and thing. The rise of the machine, as manifest in the prosthetic body part, embodied a threat to the hegemony of the physically and organically whole human.

To assess the variety of fictional responses to concepts of prosthesis and autonomy, this chapter is split into three thematic sections, which each include analyses of works from across the period that this book explores. The first section examines representations of devices that either enable or inhibit user agency. Here I focus closely on two writings that simultaneously reveal and dismiss the extent to which technological apparatuses can provide physical autonomy to their users: Edgar Allan Poe's "The Man That Was Used Up" (1839) and Herman Melville's *Moby-Dick* (1851).[19] As I show, despite presenting entirely contrasting prostheses in terms of sophistication, these sources in varying ways problematized the demand for prostheses that was encouraged by a society that privileged wholeness. Prostheses are depicted interchangeably as dehumanizing and ineffective.

A specific example of a character (Captain Boomer) who uses a non-normative prosthesis to maintain authority in *Moby-Dick* provides a segue to the next section, which explores prosthetic technologies imagined to transgressively enable their users. The dominance of physical wholeness is challenged more directly in the stories covered, which present users who find violent means with which to assert themselves. This section explores in detail the enduring motif of prosthesis as weapon, providing close readings of a variety of writings including Thomas Hood's *Miss Kilmansegg and Her Precious Leg* (1840–1841) and Arthur Conan Doyle's *The Sign of Four* (1890).[20] Thinking in further depth about threatening machine agencies, the final section explores the self-acting prostheses of Parker's *The Flying Burgermaster* and Blackton's *The Thieving Hand*, among others. Collectively, my analyses consider the extent to which prostheses and their users were imagined in ways that defied hegemonic organic bodily norms.

Productive Prostheses

In Erin O'Connor's influential study of nineteenth-century amputation and lower-limb prostheses, she argues that "[m]echanizing the amputee in order to naturalize him, the discourse of prosthesis redistributes the qualities of personhood across an economy of body and machine."[21] In an optimistic manner, she demonstrates how artificial limb makers asserted the abilities of their devices to "rehabilitate recalcitrant bodies," enabling male amputees to return to work, thereby restoring the masculinity jeopardized by losing a limb and returning him to a state of autonomy.[22] Despite invoking literary works including Charles Dickens's novels *Our Mutual Friend* (1864–1865) and *Pickwick Papers* (1836–1837), and Anthony Trollope's *He Knew He Was Right* (1869), O'Connor provides limited evidence that contemporary literary depictions of prosthetics supported the utilitarian ideal espoused elsewhere.[23] Imaginative engagements with the topic complicate the view that O'Connor addresses.

In support of the utilitarian-essentialist view of prosthetics that O'Connor identifies, a small number of fictional tales supported the enabling potential of contemporary prosthetics. For example, an 1872 sketch for *Kind Words* titled "A Romance of War" tells the story of a maimed war veteran thought dead, who thanks to the use of an artificial leg and patent arm becomes a successful grocer.[24] He is also able to remarry his wife, who thought he had been killed in action and had married another man. In the 1874 *All the Year Round* short story "Peg-Legged Bob," a maimed navvy is bought a well-made artificial leg, which, though not good enough a substitute to enable him to return to manual work, allows him to become a contractor.[25] These technophilic depictions of artificial limbs are no doubt linked to the air of optimism and achievement surrounding prostheses in the wake of the American Civil War (1861–1865), a conflict that many historians hold accountable for major developments in artificial limbs.[26] In 1864, the famous American poet and physician Oliver Wendell Holmes boasted that America "has bestowed upon you and all the world an anodyne which enables you to cut arms and legs off without hurting the patient; and when his leg is off, she has given you a true artist's limb for your cripple to walk upon, instead of the peg on which he has stumped from the days of Guy de Chauliac to those of M. Nelaton."[27]

Concerning ocular prostheses, in M. E. Francis's 1900 short story "A Rustic Argus" a man with one eye is convinced by the sister of his lover to use a prosthesis.[28] He heeds her advice and wears a prosthesis with profound effect. He even claims that it enhances his eyesight. Though clearly hyperbolic, this representation followed major developments made in artificial eyes. In the 1890s, the Snellen, or "reform" eye as it was sometimes known, was developed by the Müller family in Wiesbaden in response to ophthalmologist Hermann Snellen's call for more suitable prostheses for enucleated eye sockets to be produced. These devices of a double shell design better filled the cavity left after enucleation, removing the sunken appearance that previous designs effected.[29] In 1900, Pache and Son of Birmingham were makers to the principal hospitals in Britain and could provide the "reform eye."[30]

Simultaneously aligning with the optimistic view of prosthetics outlined in the stories above while also exposing angst regarding the potentially dehumanizing effect of a human-machine splice, Edgar Allan Poe's 1839 story "The Man That Was Used Up" is a literary source that O'Connor gives close attention to.[31] In Poe's narrative, which is framed as a kind of detective story, the narrator tries to find out the secret behind Brevet Brigadier General John A. B. C. Smith, a figure who is at once mysterious and striking in appearance. As is revealed in a grotesque dressing scene, the veteran of the Bugaboo and Kickapoo campaign is a maimed man almost entirely made up of prosthetics. Due to his impressive appearance when fully assembled, Smith is shown to use devices that not only conceal his bodily losses but appear to improve upon nature. The narrator is awestruck by Smith's appearance during their first encounter: "His head of hair would have done honor to a Brutus"; his teeth were "the most entirely even, and the most brilliantly white of all conceivable teeth"; his eyes were "of a deep hazel, exceedingly large and lustrous"; his bust was "unquestionably the finest" that the narrator ever saw; his arms were "altogether were admirably modelled"; and his lower limbs were the "*ne plus ultra* of good legs."[32] Even Smith's apparent stiffness, an early sign of his artificial composition, is viewed positively by the narrator:

> There was a primness, not to say a stiffness, in his carriage—a degree of measured, and, if I may so express it, of rectangular precision, attending his every movement, which, observed in a more diminutive figure, would have had the least little savor in the world, of affectation, pomposity or constraint, but which noticed in a gentlemen of his undoubted dimensions, was readily placed to the account of reserve, hauteur—of a commendable sense, in short, of what is due to the dignity of colossal proportion.[33]

Smith's devices are notable for their mimetic capacities but are even more impressive in terms of the way that they provide the severely injured subject both a degree of physical autonomy and an impression of grandeur.

However, as the title of the story suggests, despite the impressiveness of Poe's prostheses, the narrative is less troubled with the imagined ameliorative capacities of prosthetics than it is with the effects that a human-machine splice has ontologically. When a man is more machine than human, how does this affect his personhood? Who is in control, him or his devices? It is significant that Smith is described as "*the man that was used up*" (emphasis original), rather than "the man that was *made* up."[34] James Berkley notes that in the nineteenth century "used up" could mean "not only 'to expend' or 'to exhaust' (its normal meaning today) but also 'to debunk' or 'to critique.'"[35] The title thus describes both a man who is physically reduced and a man whose very personhood is put into question; its passive verb construction implies that the man is not a user but an object of use by someone (or something) else. Smith has been "used up" by a life of military duty and, in some ways, perhaps, by his increasing reliance on prostheses. Drawing our attention to the extent to which Poe's autonomy is compromised by his reliance on technology, O'Connor argues that "The Man That Was Used Up" celebrates "the constitutive powers of prosthetics as a means of taking them to task." "Centring on a man who has more artificial parts than organic ones," O'Connor suggests,

"Poe's story critiques the logic of prosthesis by turning it inside out."[36] Like Mr. Pinto from William Makepeace Thackeray's 1862 sensation-fiction parody "The Notch on the Ax," another character who is heavily supplemented with prostheses whose representation raises questions about who or what is in control, it is unclear in the case of Smith what is part and what is whole.[37]

Though Poe's story is by no means a realistic portrayal of nineteenth-century prostheses, precisely because the devices presented are far more sophisticated than those that were available even at the end of the century, the story does anticipate the boom in prosthesis manufacture and distribution that occurred during and after the American Civil War. Highlighting the correlation between conflict and prosthesis development, Katherine Ott reveals: "The Civil War in the United States and the Napoleonic Wars in Europe initiated the first large-scale attention to prosthetics and their design and use."[38] Applying this technological narrative of progress to "The Man That Was Used Up," Vanessa Warne notes that "Poe's story anticipates [the developments in the design, manufacture, and marketing of prosthetic body parts], accurately envisioning the commercialism, technological character, and normalizing goals of post-bellum America's response to disabled war veterans."[39] The narrator's horror at witnessing Smith being literally reassembled by his slave, Pompey, at the end of the narrative, is comically contrasted with Smith's recommendations of prosthesis makers. Smith's championing of certain brands of prosthesis, including Thomas's cork legs, Bishop's artificial arms, Pettitt's shoulders, Ducrow's bosoms, Bonfanti's palate, De L'Orme's wigs, Parmly's dentures, and Dr. Williams's sight-enabling artificial eyes, brings to mind a popular feature of nineteenth-century prosthesis treatises, publications that promised potential clients that prosthetics could enable the same degree of autonomy afforded to the nondisabled in wider society.

A familiar component of the pamphlets, books, and other advertisements published by prosthetists on both sides of the Atlantic in this period was the testimonials section, or "letters pages" as it was sometimes labelled. In this popularly deployed section, the testimonials of named or anonymous users, medical professionals, or technology experts were printed together. Occasionally these were divided into subsections based on the age, gender, profession, and disability of the writer. In all cases, these letters would laud the capabilities of the products produced by the advertised prosthetist, and occasionally they would disparage the devices of other makers. In Douglas Bly's *Description of a New, Curious, and Important*

Invention (1859), a report by A. P. Sigourney, Chairman of New York State Fair Committee, compared the maker's artificial legs to rival prosthetist Benjamin Franklin Palmer's devices: "We are unanimous in the opinion that the Leg presented by DR. BLY is the best, and that it possesses advantages over the 'Palmer Leg' very desirable to the user, and creditable to its maker."[40] In his privileging of the devices of certain makers over those made by others (i.e. he recommends Thomas's cork legs but suggests that Bishop makes superior artificial arms), Smith pre-empts the kinds of debates that would occur within the rhetorical literature published by prosthesis makers from the 1850s onwards. As Kevin J. Hayes observes: "each of General John A.B.C. Smith's prosthetics is identified with a specific brand name; put together, they virtually turn him into a walking advertisement."[41]

While Poe's narrative, on the one hand, sanguinely imagines a future in which the most physically damaged bodies can be recuperated and made whole again by prostheses, it is typical, on the other hand, of nineteenth-century prosthesis narratives in that it does not fully support prosthesis in either conceptual or practical terms. "The Man That Was Used Up" makes fun of a prostheticized physical aesthetic while challenging the contemporary preference for organic physical wholeness in a couple of ways. In practical terms, Poe's narrative critiques prostheses in two ways. First, despite their impressiveness at first glance, Smith's adornments produce a curiously rigid aesthetic, which invites curiosity: though the narrator is initially wowed by Smith's appearance, he nevertheless notices Smith's rigidity, thereby stimulating an investigation. The second practical aspect of Smith's prosthesis use, which is similarly comically explored, concerns the user's assembly process. Because these devices are not fully integrated with the subject and thus need to be removed at certain times of the day—before bed, for instance—there are periods when the prosthesis user is rendered almost entirely reliant on others. Smith's rudeness towards his slave, who helps him to attach his various prostheses, reveals a frustration at such a compromise. Though presented in a disablist grotesque fashion, which makes a spectacle of the disabled body, Poe's narrative nonetheless problematizes prostheses as complete solutions to the perceived crisis of physical loss. In conceptual terms, Poe's critique of physical wholeness is also twofold. On the one hand, the hegemony of organic completeness is challenged by a prostheticized figure, someone who is neither fully organic nor physically whole, whose appearance when fully constructed is more impressive than physically normative subjects. On the other hand, the very drive to construct an appearance of integrity is rendered absurd as we witness a subject go from a state equivalent to the titular "human bundle" of Ernest George Henham's 1897 short story, to a shape that is imposing and impressive though decidedly mechanical.[42] The falseness engendered by prostheses is taken to a hyperbolic extreme. The following analysis exposes other ways in which the notion of prostheticizing the body was challenged and the hegemony of wholeness distorted in nineteenth- and early twentieth-century prosthesis narratives.

Disabling Devices

While Smith, to a degree, is remarkably enabled by his use of prosthetics, several other representations from the period focus on the inabilities of these devices to grant users autonomy. A few years before Poe's story was published, a tale titled "Wooden Legs" (1833) appeared in the *Court Magazine and Belle Assemblée* that mocked the efficacy of sophisticated artificial legs. At the time, such devices were made at an increasing rate in Britain and America, following the success of James Pott's famous Anglesey leg, which I return to below.[43] In the story, a Napoleonic War veteran mistakenly puts on a friend's expensive patented artificial leg after a heavy night of drinking. The combination of his drunken state with a device that was earlier shown to be tricky to use results in repeated falls, creating a cruelly comical scene.[44] Similarly, Hood's popular satirical poem *Miss Kilmansegg and Her Precious Leg* mocks a pompous amputee who demands an artificial leg made of solid gold. Unsurprisingly, her device is shown to be cumbersome, heavy, and impractical. While the artificial leg in Hood's poem is undoubtedly, as Warne suggests, a symbol through which the mismanagement of money is explored, the poem also exposes contemporary disquiet about the weight of artificial legs and was used in contemporary prosthesis advertising as an example of how limb prostheses should not be constructed.[45]

Once we turn to the second half of the nineteenth century, a period heralded by popular historian Guy Woodforde as "the era of false teeth for the masses" due to technological developments that made the manufacture of these technologies cheaper, we encounter numerous stories and jokes that mocked the propensity of false teeth to malfunction.[46] Published in *All the Year Round*, "Too Hard upon My Aunt" (1863), for instance, tells the story of the narrator's aunt who suffers a mysterious "illness" after eating boiled-beef in the company of a love interest.[47] As is revealed, she is a false teeth user. Her dentures broke during the meal, hence her feigned illness. Hyperbolizing the inability of wigs to stay attached to their users' heads, the 1895 *Illustrated Chips* cartoon "The Wig Wouldn't Work Like the Natural" shows how representations of ineffective prostheses continued late into the century.[48] The cartoon depicts a comic artist prone to pulling his hair, who ends up using a wig and pulling off with a "wild, mad tug," much to his embarrassment.[49] In each of these stories, the autonomy of a subject is impacted by using an artificial body part. The very employment of these contraptions, which promised to enable users to meet emerging standards of normalcy, was therefore critiqued. These imaginaries present prosthetic devices that create rather than cure physical differences.

A famous example of a prosthetic body part that is disabling rather than enabling is Captain Ahab's ivory leg in Melville's *Moby-Dick*. This epic narrative questions the notion that prostheses provide a medical fix or cure for physical loss, instead recognizing the virtues of building a more inclusive and accessible environment. In this respect, the novel projects a fairly forward-thinking way of responding to physical difference. Though, in other regards—for instance, in the way that Ahab's abhorrent monomaniacal traits are equated with and represented physically by his disability—the novel is typically disablist in its treatment of physical difference.[50] On the positive side, Ahab is granted a degree of physical autonomy in what would normally be a particularly inaccessible workspace, thanks to numerous adaptations made to his ship, the *Pequod*. These developments compensate for the limited autonomy enabled by the amputee's ivory prosthesis, a device that further injures him, "all but pierc[ing] his groin," late in the novel.[51] Counteracting the limitations of Ahab's ivory leg, the *Pequod*'s adaptations serve as a kind of extension of the captain's prosthetic device. Ahab's leg literally attaches to the deck of the vessel, simultaneously enabling him while also revealing the limitations of his ivory leg.

The *Pequod* is both an add-on to Ahab's prosthesis and a further physical manifestation of the captain's monomaniacal obsession—the leg made from the ivory of the same species that severed his leg being a related example. Ahab's peg directly joins him to his vessel. As the sailor Ishmael, who tells the story, reveals: "Upon each side of the Pequod's quarter-deck, and pretty close to the mizen shrouds, there was an auger hole, bored about half an inch or so, into the plank. His bone leg steadied in that hole; one arm elevated, and holding by a shroud."[52] Here, then, the *Pequod* is literally pegged to Ahab—or vice versa. As we learn, like Ahab's leg, the destiny of the ship, and its crew, is firmly attached to the captain's obsessive agenda: to "dismember [his] dismemberer," the white sperm whale Moby Dick.[53] As David T. Mitchell and Sharon L. Snyder have observed: "Each of the innovations and inventions" used by Ahab on the *Pequod*, including not just the auger hole and shroud but also the "iron banister" that he grips, the winch and saddle that carries him up to the ship's rigging, and the "spare boats, and spare lines and harpoons, and spare everythings" that Ishmael exhaustively describes, are "paraded not as evidence of Ahab's resourcefulness, but as proof of the extent to which he will go to fulfill his 'singular' quest."[54] The ship is also marked by Ahab's obsessiveness. The deck is all over dented "like geological stones, with the peculiar mark of [Ahab's] walk … the foot-prints of his one unsleeping, ever-pacing thought."[55] Here, then, the leg is a direct extension of Ahab's will. Like Ahab, the *Pequod* is also said to exhibit signs of physical abnormality: "She was a ship of the old school, rather small if anything; with an old fashioned claw-footed look about her."[56] Also resembling the captain, the vessel is adorned with the teeth of sperm whales, hence why it is described as the "ivory Pequod."[57] Emphasizing the affinities between Ahab and his ship, at the end of the narrative it is even said that "the rushing Pequod … seemed the material counterpart of her monomaniac commander's soul."[58]

As we can see from these convergences between Ahab and his ship, everything on the *Pequod* is unified in its purpose—in this case to enable the captain to complete his obsessive mission. Like Brevet Brigadier General John A. B. C. Smith's prosthetic parts, which are united in their purpose (to create an impression of the veteran's robustness and dauntlessness of character), here the ship's modifications work together to enable the disabled user. As we see in Poe's narrative, Melville's novel invites attention to the limitations of prostheses while condoning a prejudiced view of disabled subjects: Poe projects Smith as an object of grotesque spectacle while Melville endorses a view that links physical disability to moral corruption. However, in terms of attitude to the dominance of physical integrity, the representation of Ahab resists the impulse to prostheticize to conceal. Instead, Ahab uses alternative—though also prosthetic—methods to enable himself physically.

A second amputee character in Melville's novel, one who also makes manifest a dubious link between physical loss and violent intentions, is Captain Boomer, another victim of Moby Dick. Like Ahab, Boomer uses a prosthetic device. Though what distinguishes Boomer is that his prosthesis is specially designed to act as a weapon. In addition to reinforcing a problematic cultural association of physical difference with violence, Boomer's depiction also ironically challenges the hegemony of physical wholeness but in a remarkably different way. Here a prosthesis user is provided with a device that extends his abilities in a focused direction. As fighting cocks were sometimes adorned with metal spurs, Boomer is provided a prosthesis that makes him better prepared for combat. He therefore challenges physical hierarchies that placed the disabled below the nondisabled. Despite the Captain's objections, it is quite apparent that violence was his intention when he ordered the device. As the captain's doctor describes it: "he ordered the carpenter to make it; he had that club-hammer there put to the end, to knock some one's brains out with, I suppose, as he tried mine once. He flies into diabolical passions sometimes."[59] Here, we learn that the captain's prosthesis is designed specially to fit not just his body but also his violent temperament. The prosthesis, then, is an extension of the captain's volatile character as much as it is a replacement for his severed arm. In a similar but not identical way to Ahab, Boomer is revealed to be mentally unhinged, once more problematically suggesting that psychological trauma is an inevitable consequence of physical loss.

But unlike Ahab, Boomer is not interested in seeking vengeance. He admits that he has seen Moby Dick twice since his arm was lost. When asked by Ahab if he could not "fasten," in other words, capture the whale, Boomer retorts, "Didn't want to try to: ain't one limb enough? What should I do without this other arm? And I'm thinking Moby Dick doesn't bite so much as he swallows." Boomer later contemplates that "[t]here would be great glory in killing him, I know that; and there is a ship-load of precious sperm in him, but, hark ye, he's best left alone."[60] It is therefore clear that unlike Ahab, Boomer's motives lie in glory and financial reward rather than revenge. Regardless of his reluctance to pursue "the white whale" and his differing motives to Ahab, Boomer's club-shaped prosthesis is nonetheless represented as a physical manifestation of his own mental instability, one that ironically empowers the amputee over his nondisabled crew.

As we learn, Boomer went against the advice of his ship's doctor, Bunger, by having such a crudely shaped device fitted. "I had no hand in shipping that ivory arm there," Bunger states, "that thing is against all rule."[61] The rules that Bunger speaks of perhaps refer to the intended concealing capacities of prosthetic devices. This commitment to the mimesis of the organic body part is underscored by George E. Marks in his

1888 treatise on the artificial limbs produced by his family's firm A. A. Marks. Marks describes the social expectations that surrounded prosthesis design in the mid-century: "The demand for artificial limbs was noticeably increasing; the field was growing larger; the cry for something more durable and more approximate to nature came from every quarter."[62] Despite these directives, some prosthesis users fashioned devices designed for violent purposes. For instance, British artificial limb and orthopaedic appliance maker Henry Heather Bigg made an artificial hand furnished with a dagger for a furs collector working for the Hudson Bay Company. The device was produced upon request as the amputee wished to be able to defend himself against wild animals.[63] In Melville's story, Bunger is clearly not a fan of Boomer's prosthesis, in part because of the captain's violent past—Bunger suggests to Ahab that the captain gave him "bowl-like cavity in his skull."[64] While Boomer strongly refutes this claim, his very denial conjures an image of violence, containing a thinly veiled threat: "Oh, you solemn rogue, you—you Bunger! was there ever such another Bunger in the watery world? Bunger, when you die, you ought to die in pickle, you dog, you should be preserved to future ages, you rascal."[65] Boomer clearly attempts to sound jocular, but his over-emphatic tone and use of morbid imagery suggests an immovable presence of violence within his imagination.

Boomer's prosthesis challenges the hegemony of the normal body in two ways: first, the club-shaped prosthesis defies social and medical expectations that prosthetic devices should look and function like real human body parts; second, in the form of his club arm, Boomer is provided a violent and immediate means with which to assert himself, which in one respect provides him with a physical advantage over his nondisabled peers. Boomer can be read to anticipate violently equipped amputees of the modern era, such as RoboCop, who has a pistol concealed in his robotic leg; Merle Dixon from the popular AMC series *The Walking Dead* (2010–), who, like Bigg's patient, has a prosthetic bayonet arm. The same might be said of Cherry Darling from Robert Rodriguez's *Planet Terror* (2007), who has an assault rifle (complete with grenade launcher) in the place of an artificial leg.[66] As the next section shows, in addition to imagining prostheses specifically designed for confrontation, quotidian prosthetics, such as wooden legs, were represented as devices that could be used in threatening ways.

Prostheses as Weapons

Since the "turn to users" that was inspired by the work of Ruth Schwartz Cowan, Science and Technology Studies (STS) scholars have shifted their attention from investigating the impact of inventors and designers to focus on the user.[67] Most recently, these scholars have focused on "how users consume, modify, domesticate, design, reconfigure, and resist technologies."[68] While we cannot be certain that fictional portrayals of prosthesis directly informed the modifications made to artificial body parts by contemporary prosthetists, we can be sure that these portrayals drew from and informed social and cultural views of prosthesis design and use. By exploring the non-normative uses of prosthetic devices in fictional accounts, I demonstrate how prosthetic body parts were imagined as devices that were not necessarily capable of restoring the appearance and function of a lost body part but were able to provide their users with deadly and close-to-hand weapons. The logic of prostheticizing the body to recomplete it was therefore complicated by literary characters who provided a powerful challenge to the hegemonic normal body.

In literary works, prosthesis use often has little to do with restoring bodily wholeness. In various examples from nineteenth- and early twentieth-century literature and film, we see prostheses fulfil various unusual and often comical purposes far removed from the ideal of prostheses as cures for physical loss. For instance, the *Satirist* published a story in 1845 that encouraged women to use wigs for making lockets to send to lovers so that they could entertain multiple partners without going bald.[69] In "The Lame Landlord's Story," a wooden leg is made into a cupboard that stocks sweetmeats, a pipe, and tobacco.[70] By comparison, in Mark Twain's *Roughing It* (1872) a vagabond sneakily uses a cork leg to obtain brandy. After an "awkward express empoyé" drops a one-hundred-pound brick on the man's foot, the victim falls to the ground clutching his apparently damaged extremity. When a "sympathizing crowd" come to the man's aid and attempt to remove his boot, he screams louder, calling for "Brandy! for Heaven's sake brandy!"[71] Having rejected the offer of having a doctor called, the vagabond asks for more brandy and is given two bottles in addition to the half pint he had already consumed. As it turns out, the man is not hurt since the brick falls upon his cork leg. The man merely pretended to be in pain to obtain liquor. Taking an entirely different direction, an 1895 cartoon for *Illustrated Chips* illustrated how well adapted peg legs were for turning fields.[72] Though not all engaging with disability as we understand it today but rather (in the case of baldness) perceived bodily non-normativity, writings such as these provide intriguing instances of what Tobin Siebers calls "disability masquerade": "an alternative method of managing social stigma through disguise, one relying not on the imitation of a dominant social role but on the assumption of an identity marked as stigmatized, marginal, or inferior."[73] The imagined prosthesis users depicted in the *Satirist* narrative, "The Lame Landlord's Story," Twain's sketch, and the *Illustrated Chips* cartoon do not attempt to pass as normal. Instead, they perform, utilize, and, in certain cases, exaggerate physical difference to achieve success in ableist environments.

In other cultural sources, alternatives to normativity were imagined in the form of prostheses that provide effective defensive capabilities. For example, in a 1904 article for *Judy*, a contributor comically extols the multitudinous tasks that a set of false teeth could perform:

> False teeth can be used for letter clips, clothes pegs, and several other useful purposes. They can be fitted with a spring and placed on the floor to act as a trap for burglars. The burglar generally creeps about noiselessly in his stockinged feet, and when he incautiously places his foot upon one of these traps, which have been left in the room or upon the stairs for that purpose, he lets out a yell which awakens the inmates and gives the and him the alarm.[74]

In a similarly comical fashion, Arthur Cooper's 1903 short film *Blind Man's Bluff* depicts a beggar with an apparent vision impairment and amputated leg who strikes a passer-by with his wooden leg after he is given a bogus coin.[75] The film not only reimagines the association between disability, prosthesis use, begging, and imposture (a topic that I discuss at length in Chap. 4), but also reveals a common trope in the representation of human prosthesis in the long nineteenth century: the portrayal of the prosthetic weapon.

Prosthetic legs used as improvised bludgeons appeared in numerous stories in the nineteenth and early twentieth centuries, including in Hood's *Miss Kilmansegg and Her Precious Leg*, Henry Clay Lewis's "The Indefatigable Bear Hunter" (1850), R. M. Ballantyne's *Why I Did Not Become a Sailor* (1864), D. B. McKean's "A Wig and a Wooden Leg" (1886), and Arthur Conan Doyle's *The Sign of Four*.[76] In 1899, the *Nottinghamshire Guardian* printed a short report on a wooden-leg user who had been charged for assaulting a railway officer with his wooden leg.[77] Similarly, in 1904 the *Daily Mail* reported another real-life account of such an incident in which an individual named John Feenan used his limb as a "shillelagh" when intoxicated.[78] In the rest of this section, I examine three of these sources—Hood's poem, Lewis's short story, and Doyle's novel—to show how this trope encompassed a variety of complex attitudes to contemporary standards of prosthetics, the human-technology relationship more broadly, and the extent to which the power of the normate could be challenged by the violent potential of lower-limb prosthetics.

To begin with Hood's poem, a writing well known throughout the nineteenth century, we see a violent representation of a golden leg that, though absurdly humorous, became a popular paradigm for how false limbs should not look or function.[79] In Hood's poem, Miss Kilmansegg is brutally murdered by her husband, an avaricious Italian count who has accumulated considerable gambling debts. The count clubs his wife to death with the very item that he wishes to possess: the countess's golden leg. While the leg displayed in Hood's poem is by no means a realistic portrayal of a nineteenth-century prosthetic device, its representation as a weapon nonetheless reveals some of the apprehensions that surrounded prosthesis design in this period. One reason that Miss Kilmansegg's leg makes such an excellent club is because of its weightiness. The countess's prosthesis is made of "Gold—solid gold throughout" and is said to be "As solid as man could make it."[80] Considering the density of gold, it is not surprising that the device is so heavy. The hefty and cumbersome nature of Miss Kilmansegg's prosthesis is revealed by the descriptions of her noisy gait:

> When slow, and heavy, and dead as a dump,
> They heard a foot begin to stump,
> Thump! lump!
> Lump! thump![81]

The noisiness of the device becomes a nuisance to Miss Kilmansegg's murderous husband, who tries to convince his wife to exchange her golden leg for a wooden one, in part because of "the unbearable thumping" that the device makes—the other reason being so that he can sell the golden leg to pay off his gambling debts.[82] Here, the efficacy of the artificial body part is questioned, since the device clearly does not enable its user to pass as someone with two natural lower limbs; its obnoxiousness both visually and audibly brings its presence to the fore. Miss Kilmansegg's leg does precisely the opposite of what good artificial legs were supposed to do: it invites rather than evades attention. The device thus works in opposition to the social demands for silent artificial limbs that do not call attention to themselves. Her leg is not only unhelpfully heavy but also so ostentatious as to attract the unwanted attention of a suitor with murderous designs.

Due to its weight and clumsy design, Miss Kilmansegg's leg is shown to be more effective as a weapon than as a functional replacement for a leg. The cumbersome weightiness of the leg, which is not articulated, compromises the mobility of the user, but makes for an excellent weapon. In several instances the device is displayed as an impractical appendage. While the stiffness of Miss Kilmansegg's prosthesis makes her leg look like that of "a *Figuranté*" at "Her Fancy Ball," practical activities, such as walking and climbing over obstacles, are made difficult by her choice of prosthesis.[83] As the speaker notes, "She hated walking in any shape,/And a country stile was an awkward scrape."[84] In contrast to these difficulties, the leg is shown to be an excellent bludgeon by her barbarous husband. The prosthesis makes short work of the hapless Miss Kilmansegg:

> 'Twas the Golden Leg!—she knew its gleam!
> And up she started, and tried to scream,—
> But ev'n in the moment she started—
> Down came the limb with a frightful smash,
> And, lost in the universal flash
> That her eyeballs made at so mortal a crash,
> The Spark, called Vital, departed![85]

This scene makes explicit the power of the artificial part. In the violent collision between flesh and gold, only the latter survives, revealing the impressionable, delicate nature of the former and the hard, uncompromising form of the latter. The prosthesis is rendered other to body by its material properties, which so drastically contrast with that of the body. The way in which the prosthetic smashes skull, haemorrhages brain, and ultimately kills so effectively—in one blow—makes manifest the discrepancy between the natural and the artificial. Here, then, the prosthetic part, a device designed to make the body whole again, does quite the opposite as it not only further damages its intended user but is used to kill her. In this way, the impulse to prostheticize, an inclination buttressed by a culturally enforced desire to present oneself as physically whole, is challenged in two ways: on the one hand, the false part is shown to be dangerously incongruous to the natural body; on the other hand, exaggerating real-life unease about the prohibitive price of top-of-the-range artificial limbs, the false part is valued more greatly than human life.

Articulating the prosthesis as a weapon in a more sanguine manner, Lewis's "The Indefatigable Bear Hunter" effectually unites the user and device. Following the loss of his leg in a hunting accident, the bear hunter who is described in the title, Mik-hoo-tah, continues to hunt against the advice of his doctor. After a hunt goes wrong and he is forced into a fist fight with a bear, Mik uses his wooden leg to beat his furry foe to death. Though brutal, such a use of a wooden leg brings about a more intimate relationship between hunter and prosthesis. The relationship between user and device does not begin on strong footing, however. Prior to and even during his fight with the bear, Mik is revealed to be rather discontented with his wooden leg. Mik did not want to lose his actual leg in the first place (despite its mangled condition): as the doctor reveals, Mik "opposed [amputation] vehemently."[86] Similarly, after the doctor convinces Mik that amputation is the only fit course of action, and the life-saving procedure is promptly performed, the hunter falls into a state of depression, losing a considerable amount of weight—the doctor remarks "I have never seen anyone fall off so fast."[87] When the doctor asks Mik if he has contracted consumption, the hunter reveals that his physiological state is tied to his disability and, possibly, to his use of a rudimentary prosthetic: "Doc, it's grief, poor sorrur, sorrur, Doc! When I looks at what I is now and what I used to be!"[88] In the heat of the battle, Mik curses his peg leg several times for getting stuck in the uncompromising swamp terrain, referring to it as an "infurnal" and "d[amne]d thing."[89] Despite his reservations, after he successfully slays the bear using his prosthesis, Mik celebrates his success. "I hollered," Mik reveals, "I had whipped a bar in a fair hand to hand fight—me, a old, sickly, one-legged bar hunter!"[90] It is significant that Mik considers the bout with the bear "a fair hand to hand fight," entirely forgetting the fact that he used a weapon (albeit an unusual one) to bludgeon his opponent. He clubs the bear so hard with his wooden leg that its "flesh giv in to the soft impresshuns of that leg," before finishing off his unconscious adversary with a second, more conventional weapon: a knife.[91] By considering the encounter a fist fight, Mik unconsciously reveals his acceptance of the prosthetic device as a part of his anatomy.

Revealing to the doctor his triumph over the bear, Mik also ironically praises "the mederkal perfeshun for having invented sich a weepun!"[92] To Mik the prosthetic limb is noticeably more effective as a weapon than it is a substitute for an organic limb. This line of thought is not surprising when one compares Mik's locomotive immobility with the devastating efficacy of his prosthesis as a club. Mik explains that he had "*only one leg that cood run!*" suggesting that his only option with the bear was to fight rather than flee.[93] Mik's frustration at being rendered static by his leg's inability to function on soft terrain is mitigated by the bear hunter's very use of the prosthetic as a bludgeon. Though we learn relatively little about the design of Mik's prosthesis—other than the fact that it is crafted by the Swamp Doctor narrator himself—we can safely assume that, since the device was not made by a prosthesis specialist, it is of a rudimentary design. This prosthetic most likely resembles a peg rather than an artificial leg. The narrator at no point claims to be an expert in prosthesis manufacture. He elsewhere reveals the extent to which doctors in his position are forced to improvise when he describes the equipment used to amputate Mik's leg. After listing implements including "[a] couple of bowie knives, one ingeniously hacked and filed into a saw, a tourniquet made of a belt and piece of stick, a gun screw converted for the time into a tenaculum, and some buckskin slips for ligatures," the Swamp Doctor provides the following message to more affluent practitioners: "The city physician may smile at this recital, but I assure him many a more difficult operation than the amputation of a leg has been performed by his humble brother in the swamp with far more simple means than those I have mentioned."[94]

Like the amputation itself, which, as shown by Mik's subsequent bout of illness and weight loss, is by no means a complete success, the doctor's prosthesis is certainly not perfect. As it is of a rudimentary, peg-like design, it is not entirely surprising that the prosthesis functions better as a club than as a false leg. Unlike real legs, peg legs were predominantly unarticulated, rigid, and occasionally rather weighty thus making them ideal instruments for bludgeoning. That the false leg is better used as a weapon than as a practical and ameliorative replacement for a real leg underlines the otherness of primitive forms of prosthesis. Furthermore, the fact that the device is used as a club, an extension of the arm that, in the words of Harvey Green, "multiplies the force of the hand and arm because the far end travels much faster than the end in one's hands," signifies that it ironically enhances the capabilities of its user.[95] The false leg is effective to Mik as a tool rather than as a body part. The device enables the user but only in a way that compensates for its inability to grant full physical functionality. Though hardly an archetype for a new technologized bodily model,

Mik's non-normative bodily image and functionality challenged the hegemony of physical wholeness as it imagines a prosthesis user that resists passing. Mik instead uses his artificial body part in a manner that arguably makes him more capable (in certain violent ways) than a nondisabled person. The wooden leg, though not an effective prosthesis in the traditional sense, is shown to be a physical extension of the character's violent persona—recalling Captain Boomer from *Moby-Dick*. Together, Mik and his wooden leg embody a robust non-normative alternative to physical wholeness. Lewis's tale imagines a scenario in which by losing a leg one can gain a deadly weapon.

With the aptness of peg legs for violent ends, it is perhaps unsurprising that Doyle's prosthesis-using villain Jonathan Small, in *The Sign of Four*, is revealed to have used his prosthesis in an act of cold-blooded murder. While Small's false leg, like Mik's, is shown to be fundamentally flawed in terms of its use on soft ground—while attempting to escape from the authorities, he is caught by Sherlock Holmes after his peg bores itself into a muddy riverbank—the efficacy of a wooden leg as a weapon is once again lauded. After his capture at the hands of Holmes, the nondisabled "calculating machine," Small reveals his history, including an incident in which he battered a prison guard to death using his wooden leg.[96] Like Mik, Small used his prosthesis as a kind of improvised club. Small explains to Holmes how he managed to escape from a prison on a tropical island. As Small recalls, having made plans with his comrade Tonga to be collected by boat from a nearby wharf, the only guard that stood between Small and freedom happened to be, in the words of the amputee man himself, "a vile Pathan who had never missed a chance of insulting and injuring me."[97] Revealing his bloodthirsty nature, Small exacted his revenge:

> I looked about for a stone to beat out his brains with, but none could I see.
> Then a queer thought came into my head and showed me where I could lay my hand on a weapon. I sat down in the darkness and unstrapped my wooden leg. With three long hops I was on him. He put his carbine to his shoulder, but I struck him full, and knocked the whole front of his skull in. You can see the split in the wood now where I hit him.[98]

While the guard in question put up somewhat less of a fight than the enraged bear that Mik bludgeons, there remain similarities between the two disabled characters' uses of false limbs. First, in both cases the wooden leg is used in an act of improvisation, when more conventional weapons could not be sought. In Mik's case, he only reverts to using his leg as a club after the bear survives being shot, having the stock of Mik's rifle broken over its head, and being hit by the severed barrel; after all alternative options are exhausted, and the barrel of his rifle is knocked out of his hand, Mik turns to his wooden leg. Similarly, Small uses his wooden leg as a weapon only after he cannot find a suitable stone to pummel the guard with. The fact that the prosthesis is a last resort is ironic since it is undoubtedly the weapon that is closest to hand. In both cases, the efficacy of the prosthetic leg as a weapon cannot be doubted. Indeed, in both accounts, those on the wrong end of the prosthesis are despatched with relative ease, revealing the discrepancy between flesh and wood. While both narratives show that the rudimentary design of peg legs make them poor replacements for human legs—as both wooden-leg users get quite literally "stuck in the mud"—here their hardness and inhumanity are drawn attention to. In this sense, then, both stories provide a subtle critique of primitive lower-limb prostheses, suggesting that they are better adapted to being used as weapons than as functional substitutes for missing legs. In this sense, these stories challenge contemporary demands for amputees to prostheticize by suggesting that peg legs, the lower-limb prostheses most affordable to the masses, could barely enable users to walk let alone pass as normal.

Despite this seemingly important message, the treatment of prosthesis as weapon is rather comical in both stories. In *The Sign of Four*, for instance, the police officer that eventually arrests Small makes this joke: "I'll take particular care that you don't club me with your wooden leg, whatever you may have done to the gentleman at the Andaman Isles."[99] Lewis's story meanwhile is laden with absurd details about rural life in the American South designed to elicit laughter from metropolitan audiences. But both tales make a serious point about the inefficacy of nineteenth-century peg legs. These representations mobilize an alternative model of physical incompleteness: the users of artificial limbs are shown to be conspicuously deficient in locomotive capacities but enhanced when it comes to hand-to-hand (or rather *leg-to-head*) combat. Wooden-leg users are imagined to use their non-normative bodies effectively in non-conventional ways, challenging the physical hierarchies that placed the conspicuously incomplete below those who were deemed physically whole.

The use of prosthesis as weapon, however, also reveals an ableist disquietude regarding the disabled user. To use terminology borrowed from

Madeleine Akrich, while murderers such as Small do not correspond with the "projected user" envisaged by the makers of prosthetic limbs, amputees and disabled people in general were subject to a number of stereotypes that often equated physical disaggregation with moral deficiency.[100] Discussing the portrayal of disability in Western literature and film in general, Jenny Morris identifies that "beauty—and goodness—are defined by the absence of disability ... ugliness—and evil—are defined by its presence."[101] In more nuanced work on the subject, Mitchell and Snyder have suggested:

> The disabled body became an important means of artistic characterization, for it allowed authors to visually privilege something amiss or "tragically flawed" in the very biology of an embodied character. While disability had historically provided an outward sign of disfavor or monstrous inhumanity, the nineteenth century shifted the emphasis to a more earthbound principle of moral decrepitude and individual malfunction.[102]

Along similar lines, underscoring once again the long-held belief in the link between physical trauma and a propensity to violence, David A. Gerber notes that "societies have long been haunted by fears of the violent potential of veterans with unpredictable mental states."[103] As James Marten identifies, statistics arising from the American Civil War supported these worries: "In Lancaster County, Pennsylvania, a postwar increase in crime was fuelled largely by returning veterans, who committed nearly half of all offenses, especially 'moral' and property crimes."[104] In this sense, then, one can consider Small's missing leg as a physical signifier for his villainy and violence. The loss of a limb is not too different from an "anomaly," an "abnormal" facial feature that Cesare Lombroso suggested indicated that an individual "may display profound moral maladjustment."[105] While Small very much embodies the villainous stereotypes that surrounded disabled people, Mik is not too far from this typecast either. Like Small, Mik displays an evident inclination towards violence and, if anything, is more bloodthirsty in his "indefatigable" hunting of bears than the opportunistic villain of Doyle's story. In one sense, then, Mik and Small are rendered further abnormal by their uses of prostheses as weapons. Their cyborgian adoption of prosthetic technologies provides a comically treated but nonetheless culturally loaded challenge to contemporary bodily hierarchies.

Self-acting Prostheses

Alongside the prosthesis-as-weapon imaginary, a related trope, which conceptualized the dangerous potential of false body parts while mustering a challenge to preferences for organic wholeness, is the portrayal of the self-acting prosthetic. Mobilizations of this trope ask the following question: What happens if prosthetics, devices imagined to be more physically robust than human flesh, become so sophisticated that they begin to possess agency of their own? Parker's narrative poem *The Flying Burgermaster* and the popular song "Cork Leg" tell almost the same basic story about a wealthy member of the European upper classes who loses a leg and purchases a false limb, which turns out to be so sophisticated that it literally walks its user to death. While a comical reviewer writing in *Judy* in 1904 joked about the jovial possibilities of self-acting prosthetics, jesting that "Smiling teeth may be purchased by the set without extra charge for the smile," here automatous devices were presented in a more threatening manner.[106] Besides Parker's poem and "Cork Leg," a number of other fictional stories imagined unruly prosthetics, including *All the Year Round*'s "Bolderoe's Widow" (1876), Allsopp Æsop's "The Wooden Leg and the Ungrateful Pensioner" (1878), *Every Week's* "A Wooden Leg That Knows a Thing or Two" (1895), Frank Crane's "Willie Westinghouse Edison Smith" cartoons "Willie Westinghouse Invents an Automatic Arm" (1904) and "Willie's 'Handshaker' Gets Papa into Trouble" (1907), and J. Stuart Blackton's *The Thieving Hand*.[107] These stories either present self-acting devices that malfunction—that is, prostheses that refuse to stop at the command of the user—or display intelligent devices, ones whose desires are often at odds with their users. By focusing on Parker's poem and Blackton's film, I argue that these representations draw from contemporary anxieties relating to technological developments while presenting powerful non-human and physically incomplete subjects, ones that challenge the normate.

In Parker's poem, following the amputation of his leg, the comically named protagonist, Wodenblock, is delighted to learn of the also aptly named "great mechanic" Turningvort's artificial leg. The "cork leg" is "of such perfection;/So firm, yet steady, that it stood,/Walked, danced, and ran, like flesh and blood."[108] Wodenblock is attracted to Turningvort's prosthesis because of the device's ability to mimic not only the appearance but also the function of a real leg. He thus has no qualms about ordering the prosthesis. He pays a rather large sum to acquire the device—enough to pay for Turningvort's daughter's marriage dower. Initially, the protagonist's new prosthesis appears to be a success. The device is so well-formed that it is said to make the Burgermaster's other leg look feeble in comparison:

Once more a biped—Wodenblock
Stands firm and steady as a rock—
Complacently the limb he eyed,
And thought the old one by its side
Look'd thin and shabby—truth to tell,
It boasted not the graceful swell
Or taper ancle of the other,
But seemed a starving younger brother.[109]

Though the device is clearly impressively made, the fact that the artificial limb makes his other look incongruous implies that there is a discrepancy between the device and user. Unlike the contemporary representations of peg legs, here the user's body is insufficient compared to the prosthesis rather than vice versa. The false part is shown to be an improvement on nature. There are hints that the desire for prostheses that can enable one to pass has gone too far. The false part fails to conform to the demands for physical wholeness, inverting preferences for the organic.

Perhaps out of fear from encountering the kind of difficulties that Mik and Small experience while using their prostheses on rough terrain, Wodenblock initially avoids "crooked lanes and allies" on his first walk using the artificial leg. His worries are soon allayed:

With smiling looks, and air confiding,
Down broad strait streets triumphant gliding.
The leg displayed no turn for kicking,
A little whirl—a gentle ticking;
Was all the fault he could descry,
And that he thought would soon pass by.[110]

However, this confidence in the limb is short-lived. When he attempts to turn around to greet a friend, his leg is shown to possess a different agenda:

He wheeled around without reflection,
Quite in the opposite direction

To that which he had just pursued;
When—as with magic power endued
A sudden jerk, a whirling thrill—
The leg no more obeys his will;
In haste, he had omitted learning
Which spring to touch in case of turning;
And prest on one of wondrous force,
To impel him on his forward course.
The act was scarce performed, when lo!
Swift as the arrow from the bow,
He felt himself compelled to fly[.][111]

Here, the Burgermaster's leg is not only self-acting but also entirely rebellious. Unwilling to heed the commands of its wearer, the leg takes on a direction of its own at a frightful pace—one not dissimilar to Wallace's "wrong trousers." The leg proceeds to drag its user through the woods and "O'er dykes, morasses, rivers, floods," leaving the wearer "Exhausted, trembling, gasping, fainting."[112] The unrelenting march of the artificial leg eventually results in Wodenblock's death. Even after the user's death, the leg compels his skeleton to keep marching. By rendering the ghost of the amputee a "slave to all eternity," the poem presents a technological refashioning of the Christian "Wandering Jew" legend, which tells the story of a Jewish man, who as a consequence of rejecting Jesus is condemned to walk the world homeless until the second coming.[113] That the prosthesis outlives its user's body once again reveals the contrast between fragile body and tough device. The life of the mechanical part is shown to outlast the human whole, primarily because it is much more resilient.

The invulnerability of the false part was ruminated on again at the turn of the century in Wells's *The Food of the Gods* (1904).[114] In this story, the farmer Mr. Skinner is eaten by a giant rat whose great size and appetite is effected from earlier eating the titular "food of the gods." When a search party is sent out for Skinner, the only part of him found is his glass eye. This motif was also present in Rudyard Kipling's short fiction "Mrs. Bathurst" (1904).[115] Here, a charred corpse is identified as the body of a navy officer named Vickery in part because his four white artificial teeth are seen "shinin' against the black" of the carcass.[116] The way in which these devices outlast their host exposes an apprehension that humans could be outlasted by a more resilient force: technology. As I show below, Parker's poem responds to various contemporary contexts, including debates concerning vitalism and the possibility of perpetual motion; the growth of the artificial limbs trade; the rise of industrial manufacture; and the development of automated machines, such as Babbage's "Difference Engine."

While Parker's poem is no doubt comical in its tone, it clearly reacts to trepidations about the splicing of a person with a thing. The technologically sophisticated object is shown to be too advanced for its user, since it demands more exercise than the human body can endure. Describing Turningvort's despair at hearing the pathologizing nature of his creation, the speaker explains:

> His curious springs, wheels, cork, and leather,
> By rarest art combined together,
> Had done their work: and tho' by him
> Perchance this superhuman limb
> Might condescend to be directed,
> It still might spurn to be subjected
> To one, upon whose depth of science
> It felt but moderate reliance[.][117]

The device's agency but lack of sympathy underscores its inhumanity. Like an unruly piece of factory machinery, the leg has "a mind of its own" but is not sophisticated enough to deviate far from its primary function. The poem also shows the leg to be far more resilient that the body of the user, exacerbating the contrast between flesh and machine. The prosthetic leg is thus depicted as an uncontrollable mechanical other. In some ways, the poem exposes cultural misgivings about scientific authority. The worry is that, in an error of judgement, a scientist could produce a self-acting mechanical device that objects to being subservient to humankind. Here the false part, one employed to enable the user to pass as normal, quashes the user's agency, making him act oddly and unpredictably.

In this regard, the poem engages with contemporary debates related to life, energy, and the possibility of perpetual motion. Building on the division of the body and soul posited by Plato, Paracelsus, Jan Baptist van Helmont, Descartes, Georg Stahl, and the Montpellier school of medicine, eighteenth- and early nineteenth-century vitalist thinkers, such as Xavier Bichat, suggested that animal and organic life could be distinguished by a contrast between the forces of life and death. Bichat made the famous assertion that "life is the sum of the forces which resist death."[118]

Vitalism, "the theory that life is generated and sustained through some form of non-mechanical force or power specific to and located in living bodies," remained an influential yet contested biological philosophy throughout the nineteenth century.[119] In his 1840s work on "the conservation of force," German surgeon and scientific philosopher Hermann von Helmholtz sought to disprove vitalist theories by demonstrating that perpetual motion—the possibility for which he claimed was implied by vitalist thinking—was not possible.[120] As Helmholtz argued, "no perpetual motion is possible ... force cannot be produced from nothing, something must be consumed."[121] Engaging with vitalism philosophy and problematizing John Locke's enduring emphasis on life "as the distinguishing character between men and machine,"[122] Parker's poem reveals a Frankensteinian dread that a scientist could, by mistake, imbue a mechanical object with vitality, creating a conspicuously non-normative and non-human life form capable of threatening those who encounter it. Moreover, Parker's poem interrogates the vitalistic binary between animal life and the mechanical world, exploring the harmful potential of science gone wrong. Even more to the point, by displaying a prosthetic device that not only self-acts but animates without the need of a "*moving force*," Parker brings to the fore angst about mechanical devices that could move perpetually.[123] We can also read Wodenblock's automatic prosthesis in relation to the real-life development of highly sophisticated mechanical devices. Contexts such as the opening of the Liverpool and Manchester Railway in 1830, Charles Babbage's continuing work in the 1820s and 1830s on the Difference Engine, and James Pott's invention of the Anglesey Leg in 1816 and its subsequent celebrity championing by the Marquess of Anglesey—a shifting into public view of artificial limbs that Vanessa Warne suggests "hints at the disquieting prominence of prosthetic legs in Victorian culture"—are all important factors.[124]

The opening of the Liverpool and Manchester railway on 15 September 1830 was both contemporaneous to the publication of *The Flying Burgermaster* and significant for bringing the human-technology relationship into public consciousness. The opening marked a new age of British transport and a celebration of the nation's technological achievements. By linking together two of Britain's most important industrial centres with a state-of-the-art technology designed to reduce journey times, the railway line promised to alter concepts of distance in the British imagination. A thousand passengers and almost a million onlookers attended the event. The crowds were already startled by the sheer speed of the new machines, which travelled at over thirty miles per hour (as fast as the quickest horses), but that was nothing compared to the shock caused by a fatal accident that marred the opening: Tory politician William Huskisson was struck by a train, causing a deadly leg break. The public display of mechanical brutality highlighted to the hordes in attendance and the millions who would read or hear about the incident afterwards what was already all too apparent to the contemporary factory worker: the new steam-powered machines were tremendously strong, unrelenting, and often outright dangerous. As much as these technologies could annihilate time and space, they could decimate the human body.[125] Parker's image of a device that is also designed to enable locomotion, which affects the leg of a powerful local political figure and displays the ability of mechanical devices to run straight through human intervention, emerges as an imaginative response to Huskisson's contemporaneous accident. The mechanical prosthesis, a device designed to effect an appearance of completeness, appears as an ironic physical threat to organic wholeness.

Alongside the emerging spectacle of the railway and the unease it evoked, the question of the ability of the machine to think, a similarly threatening concept, was raised by Babbage's invention of the Difference Engine. Babbage came up with the idea for the Difference Engine in 1820 as a means with which to mechanize the production of mathematical tables. Despite encountering many difficulties along the way, by 1832, with the help of his engineer Joseph Clement, Babbage had completed a small section of the Difference Engine, a machine that would sadly never be completed during the inventor's lifetime. Though not the first automatic machine—clocks, trains, and textile machinery all preceded it—the Difference Engine, as a concept more than a complete artefact, was significant as it brought to the fore the notion of machine intelligence. As Doron Swade writes: "[the Difference Engine] is a landmark in respect of the human activity it replaced. In the case of textile machines or trains, the human activity they replaced was physical. The 1832 engine represents an integration of machinery into psychology."[126] Babbage himself stressed that his engine could not think per se, but its base organization "reflected the nervous system's structure as revealed by anatomists of the 1830s."[127] Despite this reserved approach, others later saw the Babbage's Difference and later Analytical Engine as devices that could replace the brain, an idea poignantly reinforced by Ada Lovelace's 1843 influential notes.[128] Babbage's friend and barrister Harry Wilmot Buxton, who wrote Babbage's biography between 1872 and 1880 declared, "the marvellous pulp and fiber of a brain had been substituted by brass and iron … [. Babbage] had taught wheel work *to think*, or at least to do the office of thought."[129] In the same way that Babbage brought concepts such as the machine and volition into conversation with one another, Parker's poem, and the stories of self-acting prostheses that proceeded it, drew from what were controversial materialist views, depicting mechanical devices whose proximity to the human was all too close and whose volition was all too strong. Simultaneously mechanical subjects and supplementary objects, these devices threatened to collapse the distinction between human and thing, part and whole.

A further important context for Parker's representation was the real-life development of artificial legs. As David M. Turner and Alun Withey discuss in their article on eighteenth-century technologies of the body, as early as the sixteenth century there emerged "a growing division between devices that were strictly functional, such as the peg legs doled out to poor amputee, and more sophisticated prosthetics that not only restored movement but actually resembled the missing body part."[130] Pioneers of early sophisticated artificial limbs included Ambroise Paré. By the late eighteenth century, prosthetists such as Monsieur Laurent of Bouchain, Thomas Ranby Reid, and Thomas Mann were receiving high praise in newspapers for the high build quality and lifelike nature of their devices. However, it was not until after the Napoleonic Wars were over that artificial-limb developments became mainstream knowledge. The device that caught public attention was that worn by the First Marquess of Anglesey, a limb that became affectionately known as the Anglesey leg.

After Henry William Paget, at the time styled as the Earl of Uxbridge, lost his leg to grapeshot in the Battle of Waterloo in 1815, he turned to Chelsea Hospital prosthetist James Potts for a device that could stand in for his lost limb. Potts, who was the first man to patent a wooden leg articulated at the knee, ankle, and toe joints in 1805, produced an artificial leg crafted to an elegant and lifelike form. It contrasted the rudimentary shape of peg legs, which, despite the early development of artificial legs, remained the preferred type of lower-limb prosthesis up until this point. As Paul Youngquist writes: "Potts's work was top-notch, and he created a noble leg for a noble patriot."[131] His device, which came to be known as the Anglesey leg, served as a model for the development of numerous patented artificial limbs as the trade ascended in Europe and America over the course of the nineteenth century. At a time when England was reportedly replete with amputees during the aftermath of the Napoleonic Wars

(1803–1815),[132] *The Flying Burgermaster* responded to the issue of rehabilitation by imagining a hyperbolized version of a prosthesis like the Anglesey leg as a technology so astonishingly sophisticated that it could override the will of its user.

Drawing from similar contexts to *The Flying Burgermaster*, appearing first in the 1820s or 1830s, the oft-repeated popular "Cork Leg" narrative was also written in a comical style, and told more or less the same story as Parker's poem. A poetic version has been attributed by many to Thomas Hudson, with potentially the earliest (though anonymous) recorded version of it dated between 1819 and 1844 in the Bodleian Library.[133] However, a prose version by Henry G. Bell, entitled "The Marvellous History of Mynheer von Wodenblock," was published in *Edinburgh Literary Journal* in 1829 and subsequently reprinted in the same author's book *My Old Portfolio; or Tales and Sketches* in 1832.[134] The story was so popular that it was set to music by Jonathan Blewitt around 1830 (see Fig. 3.1) and was reprinted numerous times in both verse and prose form. Like Parker's poem, the various incarnations of "Cork Leg" drew from contemporary trepidation about machine volition. In the version published in Ashton's collection, like Wodenblock, Mynheer von Clam, the amputated protagonist of "Cork Leg," puts faith in the mimetic capacities of technology, declaring "on two crutches I'll never stalk,/For I'll have a beautiful leg of cork."[135] In this poem we get an even greater sense of the sophistication of the prosthesis:

> Each joint was as strong as an iron beam,
> The springs a compound of clockwork and steam.
> The fine shape gave Mynheer delight,
> And he fixed it on and screwed it tight.[136]

In the same manner that Wodenblock and John A. B. C. Smith attach their limbs, Mynheer screws his leg on. This mechanical mode of attaching the prosthesis emphasizes its non-humanness, making its ability to self-act even more disconcerting. The act of screwing is very much associated with the joining together of fabricated objects for the purposes of construction. Yet, as Green points out, screws were rarely used in the nineteenth century except in large industrial contexts such as factories.[137]

This fact brings to the fore a link between these narratives and anxieties regarding the factory system. Humans tend not to be threaded thus this act suggests a process of mechanization on the part of the prosthesis user.

Fig. 3.1 A cartoon etched by Joe Lisle to advertise "The Cork Leg," set to music by the composer Jonathan Blewitt. Joe Lisle after Jonathan Blewitt, *Six Scenes Narrating the Fate of a Cork Leg, the Invention of a Dutch Artist*, c. 1830, etching with engraving, Wellcome Collection, London. Image courtesy of the Wellcome Collection. https://wellcomecollection.org/works/gfb88fqh.

These depictions thus echo Karl Marx's warning, expressed most famously in *Capital* (1867), that the factory system renders humans a mere appendage of the machine at which they work.[138] "The Cork Leg" drew from contemporary apprehensions about the human-machine relationship while bringing into focus a comically pitched but nonetheless topically profound challenge to human agency. The unruly prosthetic body part threatens the agency and health of those who attempt to imitate bodily norms. The powerful prosthetic, artificial and partial, challenges the dominance of organic wholeness by asserting an unwillingness to allow its user simply to pass as normal. Instead, it effects an agenda of its own.

By the early twentieth century, following over half a century of developments in prosthetic limbs, we see the self-acting prosthesis reimagined in equally comical yet voguishly sinister terms. In *The Cinema of Isolation*, Martin F. Norden shows how early twentieth-century film frequently draws on nineteenth-century tropes and stereotypes, using issues related to disability as exploitable material.[139] As I show, motion pictures from this period also rework narratives that engage with extant concerns about human-machine relationships, using the visual medium of film to provoke new responses regarding self-acting limbs. Moreover, by engaging with the well-established hegemony of organic physical wholeness, these depictions suggest the idea that this dominance could be challenged by increasingly intelligent and sophisticated technologies.

In *The Thieving Hand*, a 1908 film directed and produced by J. Stuart Blackton, another self-acting prosthesis is portrayed. Once again in the comic mode, here a false arm is imagined that displays an insatiable propensity to steal. The arm is bought as a reward for an honest street beggar who returns a dropped watch to a philanthropic passer-by. Amusingly, the arm is purchased in a high-street artificial limb shop, hyperbolizing the increased commercialization and availability of these products in the early twentieth century. Both the homeless man and the wealthy buyer of the limb are enthralled by the sophisticated arm that they are shown, which not only looks like an organic human arm but also moves like one after it has been wound up using a removable crank handle—a practice that reflects the way that petrol-powered automobile engines were started in this period.[140] Most miraculously, the arm moves independently from a user's body. However, as soon as the limb is purchased, for what appears a large sum, the arm squirms uncontrollably and soon starts to rob anyone and anything within sight. The arm initially steals a handkerchief from the generous purchaser of the limb, but when it attempts to remove the gentleman's tie, the user stops it. On the street, the limb robs several passers-by, who in return show scorn towards the unwitting beggar. Tired of being branded a thief, the tramp sells his arm to a pawnbroker. In the pawnbroker's shop, the arm not only attacks the owner but also steals all it can from the shop's display before crawling back and reattaching itself to its original user. Furious at finding the false limb and his most valuable items missing, the pawnbroker tells a police officer and together the pair track down and arrest the innocent user. The homeless man is then placed in a communal jail cell, where his false arm attaches itself to a criminal who also happens to be an arm amputee.

There are several obvious similarities between Blackton's film and a nineteenth-century street ballad that was printed alongside "Cork Leg" in John Ashton's 1888 collection. In "The Thief's Arm" (1888), an amputee war veteran is fitted with the arm of a deceased criminal.[141] Like the false arm in *The Thieving Hand*, this limb turns out to have an agency of its own. While attached to the body of the veteran, it unrelentingly steals wherever the subject goes. After the war veteran is arrested and executed for his arm's crimes, the arm rises and joins "a body-snatching knave,/ Who stole his master out of his grave."[142] Though the arm depicted in this poem is not mechanical, it nonetheless acts in a very similar way to the limb portrayed in *The Thieving Hand*. It is possible that the poem provided inspiration for the short film. Another possible influence was the popular song "The Steam Arm" (*c*.1835), which Kirstie Blair considers as a proto-steampunk narrative.[143] While "The Thief's Arm" reveals fear of a morally corrupt medical profession—one which recklessly transplants arms and digs up the bodies of deceased paupers for experimentation—Blackton's film presents a more nuanced satire of the prosthesis industry, which by the early twentieth century was well-established and truly international.

The visual effects used by Blackton in *The Thieving Hand* complicate the distinct binary between human body and false part that is made manifest in the earlier representations of self-acting prostheses. For its time, Blackton's film is visually impressive, providing some very convincing stop-motion animation in places. In other instances, a real arm poses as the prosthetic, in practical terms reversing the primary function of a false limb. Here, a real limb stands in for a false one, rather than vice versa. This visual blurring of organic and artificial limb not only serves a practical purpose—the use of a real arm enables a convincing deception—but also makes the mechanical prosthesis look more intelligent and potentially more threatening than the previous self-acting devices. While the earlier portrayals of self-acting artificial legs tirelessly repeat the same stock motion repeatedly, here the "thieving hand" shows initiative and adaptability as it performs more than just one task—it fights the pawnbroker, strokes the same gentleman's head, escapes from a shop, drags itself along the floor, attaches and reattaches itself to two different users, and employs a range of tactics and techniques to steal from various people. We can thus consider this device as unruly yet intelligent. The prosthesis disrupts not only the life of its user but also the lives of a great number of other people, including those whom it steals from. Since it truly does possess a mind of its own, this prosthesis provides a threat to the normative order of human society. The film, in a light-hearted fashion, brings to the fore a cinematic trope of conflict between human and machine, a theme that continues to drive modern narratives—see, for instance, Alex Proyas's *I, Robot* (2004) or any of the films from the *Terminator* franchise.[144] To this day, we remain haunted by the spectre of machine agency.

Covering a wide array of writings from across the period that this book investigates, this chapter has exposed the various ways in which literary narratives conceptualized how human autonomy was affected by prosthesis use. While the literary works from this period responded to many distinct historical factors, including international conflicts, technological developments, and shifting understandings of personhood, what binds the representations discussed here is that they each interrogate the hegemony of physical wholeness. Though often tinged by ableism, the representations discussed here either expose the process of prostheticizing as flawed or present powerful non-normative challenges to contemporary bodily ideals. The autonomy of the prosthetic part and the human-machine hybrid are presented as threats to the normative organic body, even though both are products of ableist pressures that encouraged prostheses for passing. In the context of the literary narratives discussed, the concept of prosthesis as a disciplinary tool used to enforce and maintain bodily standards collapses as it brings about new threatening and conspicuously non-normative agencies. To borrow a phrase from literary historian Patricia Murphy, the supremacy of the physically complete body was being "attenuated even as [it was] being valorised."[145]

Notes

1. Nick Park, *Wallace and Gromit: The Wrong Trousers*, *BBC*, 1993.
2. Erin O'Connor, *Raw Material: Producing Pathology in Victorian Culture* (Durham: Duke University Press, 2000), 106.

3. Tamara Ketabgian, *The Lives of Machines: The Industrial Imaginary in Victorian Literature and Culture* (Ann Arbor: University of Michigan Press, 2011), 2.
4. [Frances Parker], the Countess of Morley, *The Flying Burgermaster: A Legend of the Black Forest* (n.p.: F. Morley, 1832); [Thomas Hudson,?],"Cork Leg," in *Modern Street Ballads*, ed. John Ashton (London: Chatto & Windus, 1888); J. Stuart Blackton, *The Thieving Hand*, Vitagraph Company of America, 1908, YouTube, accessed June 16, 2018, https://www.youtube.com/watch?v=J2Dr0MBWXgQ.
5. Jacques Derrida, "That Dangerous Supplement," in *Of Grammatology*, by Jacques Derrida (Baltimore: John Hopkins University Press, 1997), 144.
6. Ketabgian, *The Lives*, 19.
7. See Laura Otis, *Networking: Communicating with Bodies and Machines in the Nineteenth Century* (Ann Arbor: University of Michigan Press, 2001), 9–10; Nicholas Daly, *Literature, Technology, and Modernity, 1860–2000* (Cambridge: Cambridge University Press, 2004), 10–33; Martin Willis, *Mesmerists, Monsters, and Machines: Science Fiction and the Cultures of Science in the Nineteenth Century* (Kent: Kent State University Press, 2006), 98–114; Herbert Sussman, *Victorian Technology: Invention, Innovation, and the Rise of the Machine* (Santa Barbara: ABC-CLIO, 2009), 39–51; Ketabgian, *The Lives*, 51–54; and Roger Smith, *Free Will and the Human Sciences in Britain, 1870–1910* (London: Pickering & Chatto, 2013), 1–8.
8. Thomas Henry Huxley, "On the Hypothesis That Animals Are Automata and Its History," *Nature* 10 (1874).
9. Andrew Ure, *The Philosophy of Manufactures; or, an Exposition of the Scientific, Moral and Commercial Economy of the Factory System of Great Britain* (London: C. Knight, 1835), 367.
10. Charles Babbage, *An Essay on the General Principles Which Regulate the Application of Machinery to Manufactures and the Mechanical Arts* (London: W. Clowes, 1827), 6.
11. Julien Offray de La Mettrie, *L'Homme Machine* (Paris: F. Henry, 1865); Jenny Uglow, "Introduction: 'Possibility,'" in *Cultural Babbage: Technology, Time and Invention*, ed. Francis Spufford and Jenny Uglow (London: Faber and Faber, 1997), 16.
12. Anson Rabinbach, *The Human Motor: Energy, Fatigue, and the Origins of Modernity* (Berkeley: University of California Press, 1990), 2.
13. See Lorraine J. Daston, "British Responses to Psycho-Physiology, 1860–1900," *Isis* 69, no. 2 (1978); Sally Shuttleworth and Jenny Bourne Taylor, eds., *Embodied Selves: An Anthology of Psychological Texts 1830–1890* (Oxford: Oxford University Press, 1998); Rick Rylance,

Victorian Psychology and British Culture: 1850–1880 (Oxford: Oxford University Press, 2000); and Smith, *Free Will.*
14. Mary Shelley, *Frankenstein, or the Modern Prometheus* (London: Norton, 1996); Daniel Dinello, *Science Fiction Visions of Posthuman Technology* (Austin: University of Texas Press, 2005), 41.
15. Daly, *Literature*, 1; Dion Boucicault, *After Dark: A Drama of London Life in 1868, in Four Acts* (New York: De Witt Publishing House, 1868).
16. John Plunkett et al., eds., *Victorian Literature: A Sourcebook* (Basingstoke: Palgrave Macmillan, 2011), 274.
17. Charles Dickens, *Dombey and Son* (London: Penguin, 2002); Charles Dickens, "The Signal Man," *All the Year Round* 16, no. 400 (1866).
18. H. G. Wells, *The Time Machine* (Oxford: Oxford University Press, 2017); H. G. Wells, *The Island of Dr. Moreau* (London: Broadview, 2009); H. G. Wells, *The Invisible Man* (London: Penguin, 2005); H. G. Wells, *When the Sleeper Wakes*, Project Gutenberg, last modified September 17, 2016, http://www.gutenberg.org/files/775/775-h/775-h.htm.
19. Edgar Allan Poe, "The Man That Was Used Up: A Tale of the Late Bugaboo and Kickapoo Campaign," in *The Works of the Late Edgar Allan Poe*, by Edgar Allan Poe, vol. 4 (New York: Blakeman & Mason, 1859); Herman Melville, *Moby-Dick* (London: Penguin, 2012).
20. Thomas Hood, *Miss Kilmansegg and her Precious Leg*. In *Selected Poems of Thomas Hood*, ed. John Clubbe (Cambridge: Harvard University Press, 1970); Arthur Conan Doyle, *The Sign of Four* (Garches: Feedbooks, n.d.).
21. O'Connor, *Raw Material*, 123.
22. Ibid., 125.
23. Charles Dickens, *The Pickwick Papers* (New York: Hurd and Houghton, 1870). Charles Dickens, *Our Mutual Friend* (London: Everyman, 2000); Anthony Trollope, *He Knew He Was Right* (Oxford: Oxford University Press, 2008).
24. "A Romance of War," *Kind Words*, May 1, 1872: 145.
25. "Peg-Legged Bob," *All the Year Round* 11, no. 280 (1874).
26. See, for instance, Lisa Herschbach, "Prosthetic Reconstruction: Making the Industry, Re-Making the Body, Modelling the Nation," *History Workshop Journal* 44 (1997); David D. Yuan, "Disfigurement and Reconstruction in Oliver Wendell Holmes's 'The Human Wheel, Its Spokes and Felloes,'" in *The Body and Physical Difference: Discourses of Disability*, eds. David T. Mitchell and Sharon L. Snyder (Ann Arbor: University of Michigan Press, 1997); Stephen Mihm, "'A Limb Which Shall Be Presentable in Polite Society': Prosthetic Technologies in the Nineteenth Century," in *Artificial Parts, Practical Lives: Modern Histories of Prosthetics*, eds. Katherine Ott, David Serlin, and Stephen Mihm (New York: New York University Press, 2002); and Guy R. Hasegawa, *Mending*

Broken Soldiers: The Union and Confederate Programs to Supply Artificial Limbs (Carbondale: Southern Illinois University Press, 2012).

27. Oliver Wendell Holmes, "The Human Wheel, Its Spokes and Felloes," *Sounds from the Atlantic*, by Oliver Wendell Holmes (Boston: Ticknor and Fields, 1864), 324.
28. M. E. Francis, "A Rustic Argus," *Longman's Magazine* 35, no. 210 (1900).
29. Katherine Ott, "Hard Wear and Soft Tissue: Craft and Commerce in Artificial Eyes," in *Artificial Parts, Practical Lives: Modern Histories of Prosthetics*, eds. Katherine Ott, David Serlin, and Stephen Mihm (New York: New York University Press, 2002), 153–54. Also see I. den Tonkelaar, H. E. Henkes, and G. K. Van Leersum, "Herman Snellen (1834–1908) and Müller's 'Reform-Auge': A Short History of the Artificial Eye," *History of Ophthalmology* 4 (1991).
30. "Nineteenth- and Twentieth-Century Prostheses," The College of Optometrists, accessed June 20, 2018, https://www.college-optometrists.org/the-college/museum/online-exhibitions/virtual-artificial-eyes-gallery/19th-and-20th-century-prostheses.html.
31. O'Connor, *Raw Material*, 132–35.
32. Poe, "The Man," 315–16.
33. Ibid., 317.
34. Ibid., 325.
35. James Berkley, "Post-Human Mimesis and the Debunked Machine: Reading Environmental Appropriation in Poe's 'Maelzel's Chess-Player' and 'The Man That Was Used Up,'" *Comparative Literature Studies* 41, no. 3 (2004): 372.
36. O'Connor, *Raw Material*, 132.
37. William Makepeace Thackeray, "The Notch on the Ax—a Story a La Mode," *Roundabout Papers*, by William Makepeace Thackeray (Sydney: ReadHowYouWant, 2009).
38. Katherine Ott, "The Sum of Its Parts: An Introduction to Modern Prosthetics," in *Artificial Parts, Practical Lives: Modern Histories of Prosthetics*, ed. Katherine Ott, David Serlin, and Stephen Mihm (New York: New York University Press, 2002), 26.
39. Vanessa Warne, "'If You Should Ever Want an Arm': Disability and Dependency in Edgar Allan Poe's 'The Man That Was Used Up,'" *Atenea* 25, no. 1 (2005): 104.
40. Douglas Bly, *A Description of a New, Curious, and Important Invention* (Rochester: Rochester Evening Express, 1859), 6.
41. Kevin J. Hayes, "Visual Culture and the Word in Edgar Allan Poe's 'The Man of the Crowd,'" *Nineteenth-Century Literature* 56, no. 4 (2002): 458.

42. Ernest G. Henham, "A Human Bundle," *Temple Bar* 111, no. 438 (1897). See the intro to Chapter 2 and Ryan Sweet, "'A Human Bundle': The Disaggregated Other at the *Fin de Siècle*," *Victorian Review* 40, no. 1 (2014).
43. "Wooden Legs," *The Court Magazine and Belle Assemblee* 2, no. 3 (1833).
44. Such cruel humour mirrors Simon Dickie's observations about physical difference remaining an accepted topic for jests up until the mid-nineteenth century. Simon Dickie, "Hilarity and Pitilessness in the Mid-Eighteenth Century: English Jestbook Humor," *Eighteenth-Century Studies* 37, no. 1 (2003): 18.
45. Vanessa Warne, "'To Invest a Cripple with Peculiar Interest': Artificial Legs and Upper-Class Amputees at Mid-Century," *Victorian Review* 35, no. 2 (2009); Ryan Sweet, "'Get the Best Article in the Market': Prostheses for Women in Nineteenth-Century Literature and Commerce," in *Rethinking Modern Histories of Prostheses in Anglo-American Commodity Cultures, 1850–1960*, ed. Claire L. Jones (Manchester: Manchester University Press, 2017).
46. John Woodforde, *The Strange Story of False Teeth* (London: Routledge and Kegan Paul, 1968), 87–92.
47. "Too Hard Upon My Aunt," *All the Year Round* 10, no. 242 (1863).
48. "The Wig Wouldn't Work like the Natural," *Illustrated Chips* 10, no. 248 (1895): 3.
49. Ibid., 3.
50. In terms of Ahab's representation, Mitchell and Snyder write, "Ahab's dismemberment and 'incomplete' physicality—now simulated with a whalebone substitute—supplies Melville's characterization with both a personal motive and an identifying physical mark. These two aspects function in the novel as a deterministic shorthand device for signifying the meaning of Ahab's being." David T. Mitchell, and Sharon L. Snyder, *Narrative Prosthesis: Disability and the Dependencies of Discourse* (Ann Arbor: University of Michigan Press, 2000), 123.
51. Melville, *Moby-Dick*, 538.
52. Ibid., 144.
53. Ibid., 197.
54. Ibid., 147, 114; Mitchell and Snyder, *Narrative Prosthesis*, 122.
55. Melville, *Moby-Dick*, 187.
56. Ibid., 82.
57. Ibid., 270.
58. Ibid., 492.
59. Ibid., 511.
60. Ibid., 512–13.
61. Ibid., 511.

62. George E. Marks, *Marks' Patent Artificial Limbs with Rubber Hands and Feet* (New York: A. A. Marks, 1888), 9.
63. Henry Heather Bigg, *On Artificial Limbs, Their Construction and Application* (London: John Churchill, 1855), 73.
64. Melville, *Moby-Dick*, 511.
65. Ibid., 511.
66. *The Walking Dead*, AMC, 2010–; *Planet Terror*, directed by Robert Rodriguez (2007; Toronto: Entertainment One, 2008), DVD. There are numerous other examples of weaponized prosthetics in late twentieth and early twenty-first-century culture. For instance, we can add to this list Ash's chainsaw hand in *Evil Dead II*, directed by Sami Raimi (1987; Issy-les-Moulineaux: Studiocanal, 2019), DVD; Big Louie's meat cleaver hand attachments in *UHF*, directed by Jay Levey (1989; Beverly Hills: MGM, 2002), DVD; the minigun arm of the protagonist of *The Machine Gun Girl*, directed by Noboru Iguchi (2008; New York: Tokyo Shock, 2011), DVD; and Gazelle's blade legs in *Kingsman: The Secret Service*, directed by Matthew Vaughn (2014; Beverly Hills: Twentieth Century Fox Home Entertainment, 2015), DVD.
67. Nelly Oudshoorn, and Trevor Pinch, eds., *How Users Matter: The Co-Construction of Users and Technology* (London: Massachusetts Institute of Technology Press, 2003), 4; Ruth Schwartz Cowan, "The Industrial Revolution in the Home: Household Technology and Social Change in the Twentieth Century," *Technology and Culture* 17 (1976).
68. Oudshoorn and Pinch, *How Users*, 1.
69. "The Use of a Wig—a Hint to the Ladies," *The Satirist; or, the Censor of the Times* 675 (1845): 90.
70. "The Lame Landlord's Story," *Temple Bar* 20 (1867): 131.
71. Mark Twain, *Roughing It* (London: Penguin, 1988), 403.
72. "How He Turned It with His Wooden Leg," *Illustrated Chips* 258 (1895): 4–5.
73. Tobin Siebers, "Disability as Masquerade," *Literature and Medicine* 23, no. 1 (2004): 5.
74. "The Advantages of Having False Teeth," *Judy*, January 27, 1904, 664.
75. *Blind Man's Bluff*, directed by Arthur Cooper (1903; Alpha Trading Company).
76. Henry Clay Lewis, "The Indefatigable Bear Hunter" in *Louisiana Swamp Doctor: The Writings of Henry Clay Lewis*, ed. John Q. Anderson (Baton Rouge: Louisiana State University Press, 1962); R. M. Ballantyne, *Why I Did Not Become a Sailor*, in *Freak on the Fells; or Three Months' Rustication and Why I Did Not Become a Sailor*, by Robert Michael Ballantyne (Philadelphia: The John Winston Co., 1864); D. B. McKean, "A Wig and a Wooden Leg," *Chatterbox* 4 (1886): 27;

77. "New Use for a Wooden Leg," *Nottinghamshire Guardian*, June 17, 1899.
78. "Wooden Leg as Weapon," *Daily Mail*, January 12, 1904, 3. A similar scene was also depicted in Cooper's *Blind Man's Bluff.*
79. For more on *Miss Kilmansegg*'s popularity, see Warne, "To Invest," 95–96. For more on the poem's message about prostheses to avoid, see Sweet, "Get the Best."
80. Hood, *Miss Kilmansegg*, 782 and 785.
81. Ibid., 1042–45.
82. Ibid., 2211.
83. Ibid., 944–1224.
84. Ibid., 1933–34.
85. Ibid., 2353–59.
86. Lewis, "The Indefatigable," 235.
87. Ibid., 238.
88. Ibid., 238.
89. Ibid., 242.
90. Ibid., 244.
91. Ibid.
92. Ibid.
93. Ibid., 241.
94. Ibid., 236.
95. Harvey Green, *Wood: Craft, Culture, History* (London: Penguin, 2007), 316.
96. Doyle, *The Sign*, 13.
97. Ibid., 100.
98. Ibid., 100–101.
99. Ibid., 103.
100. Madeleine Akrich, "The De-Scription of Technical Objects," in *Shaping Technology/Building Society: Studies in Sociotechnical Change*, eds. Wiebe E. Bijker and John Law (Cambridge: Massachusetts Institute of Technology Press, 1992), 209. "Projected users" are created by and shape the work of designers of technology. As Christina Lindsay describes, "Projected users are defined with specific tastes, competencies, motives, aspirations, and political prejudices." "The innovators," Lindsay explains, "then inscribe this vision or script about the world and about the users into the technical content of the object, and thus attempt to predetermine, or prescribe, the settings the users are asked to imagine for a particular piece of technology. The prediction about the user is thus built into, or scripted into, the technology." Christina Lindsay, "From the Shadows: Users as Designers, Producers, Marketers, Distributors, and Technical Support," in *How Users Matter: The Co-Construction of Users*

and Technology, ed. by Nelly Oudshoorn and Trevor Pinch (London: Massachusetts Institute of Technology Press, 2003), 31.
101. Jenny Morris, *Pride Against Prejudice: Transforming Attitudes to Disability* (London: The Women's Press, 1996), 93.
102. Mitchell and Snyder, *Narrative Prosthesis*, 135.
103. David A. Gerber, ed., *Disabled Veterans in History*, enlarged and revised ed. (Ann Arbor: University of Michigan Press, 2012), 36.
104. Critically, though, Marten highlights a correlation between alcoholism and crime within veteran populations, suggesting that a combination of exposure to trauma and numerous social and economic conditions drove many to drink, which in turn exacerbated crime statistics. James Marten, "Nomads in Blue: Disabled Veterans and Alcohol at the National Home," in *Disabled Veterans in History*, ed. David A. Gerber, enlarged and revised ed. (Ann Arbor: University of Michigan Press, 2012), 325.
105. Cesare Lombroso, *Criminal Man*, ed. and trans. Mary Gibson and Nicole Hahn Rafter (Durham: Duke University Press, 2006), 164.
106. "The Advantages," 664.
107. "Bolderoe's Widow," *All the Year Round* 17, no. 413 (1876); Allsopp Æsop, "The Wooden Leg and the Ungrateful Pensioner," *Funny Folks* 177 (1878): 126; "A Wooden Leg That Knows a Thing or Two," *Every Week: A Journal of Entertaining Literature* 53, no. 1355 (1895): 21; Frank Crane, "Willie Westinghouse Invents an Automatic Arm," *The Wichita Daily Eagle*, 10 April 1904; Frank Crane, "Willie's 'Handshaker' Gets Papa into Trouble," *The San Francisco Call*, 7 April 1907.
108. [Parker], *The Flying*, no lines.
109. Ibid.
110. Ibid.
111. Ibid.
112. Ibid.
113. Ibid. For more on this legend and its reception history, see George Kumler Anderson, *The Legend of the Wandering Jew* (Providence: Brown University Press, 1965). A similar refashioning of this legend appeared in *Chambers's Journal*. "The Wandering Jew," *Chambers's Journal of Popular Literature, Science and Arts* 9, no. 458 (1862).
114. H. G. Wells, *The Food of the Gods and How It Came to Earth*, Project Gutenberg, last modified August 14, 2010, http://www.gutenberg.org/cache/epub/11696/pg11696-images.html.
115. Rudyard Kipling, "Mrs. Bathurst," in Rudyard Kipling, *Traffics and Discoveries* (New York: Charles Scribner's Sons, 1909).
116. Ibid., 407.
117. [Parker], *The Flying*, no lines.
118. Qtd. and trans. in A. S. Weber, ed., *Nineteenth-Century Science: A Selection of Original Texts* (Peterborough: Broadview, 2000), 7.

119. Catherine Packham, *Eighteenth-Century Vitalism: Bodies, Culture, Politics* (Basingstoke: Palgrave Macmillan, 2012), 1.
120. Weber, *Nineteenth-Century*, 278–79.
121. Hermann von Helmholtz, "On the Conservation of Force," in *Nineteenth Century Science: A Selection of Original Texts,* ed. A. S Weber, trans. E. Atkinson et al. (Peterborough: Broadview, 2000), 298.
122. Packham, *Eighteenth-Century*, 16.
123. Helmholtz, "On the Conservation," 282.
124. Warne, "To Invest," 29.
125. See Rebecca Solnit, "The Annihilation of Time and Space," *New England Review* 24, no. 1 (2003).
126. Doron Swade, "'It will not slice a pineapple': Babbage, Miracles and Machines," in *Cultural Babbage: Technology, Time and Invention*, ed. Francis Spufford and Jenny Uglow (London: Faber and Faber, 1997), 47.
127. Otis, *Networking*, 33.
128. L. F. Menabrea, "Sketch of the Analytical Engine Invented by Charles Babbage," trans. and notes Ada Lovelace, *Scientific Memoirs* 3 (1843), 666–731.
129. Harry Wilmot Buxton, *Memoir of the Life and Labours of the Late Charles Babbage Esq., F.R.S.* (Cambridge: Massachusetts Institute of Technology Press, 1988), 48–49; qtd. in Otis, *Networking*, 33.
130. David M. Turner and Alun Withey, "Technologies of the Body: Polite Consumption and the Correction of Deformity in Eighteenth-Century England," *History* 99, no. 338 (2014): 784.
131. Paul Youngquist, *Monstrosities: Bodies and British Romanticism* (Minneapolis: Minnesota University Press, 2003), 184.
132. Adrienne E. Gavin, "Dickens, Wegg, and Wooden Legs," *Our Mutual Friend*: The Scholarly Pages, accessed June 18, 2018, http://omf.ucsc.edu/london-1865/victorian-city/wooden-legs.html.
133. [Thomas Hudson], "The Cork Leg," [1819–1844] Broadside Ballads Online, accessed June 21, 2018, http://ballads.bodleian.ox.ac.uk/view/edition/5413.
134. Henry G. Bell, "The Marvellous History of Mynheer Von Wodenblock," *The Edinburgh Literary Journal, or, Weekly Register of Criticism and Belles Lettres*, no. 47 (1829): 248–49; Henry G. Bell, *My Old Portfolio; or Tales and Sketches* (London: Smith, Elder, and Co., 1832); Ray Girvan, "HG Bell: The First Cyborged Leg?" JSBlog – Journal of a Southern Bookreader, last modified July 6, 2011, http://jsbookreader.blogspot.com/2011/07/hg-bell-first-cyborged-leg.html.
135. [Hudson], "Cork Leg," no lines.
136. Ibid.
137. Green, *Wood*, 65 and 118–19.

138. Karl Marx, *Capital: A Critical Analysis of Capitalist Production*, trans. Samuel Moore and Edward Aveling (London: Kimble & Bradford, 1938). Also see Ketabgian, *The Lives*, 19–29.
139. Martin F. Norden, *The Cinema of Isolation: A History of Physical Disability in the Movies* (New Brunswick: Rutgers University Press, 1994).
140. See Joseph J. Corn, *User Unfriendly: Consumer Struggles with Personal Technologies, from Clocks and Sewing Machines to Cars and Computers* (Baltimore: John Hopkins University Press, 2011), 94–95.
141. "The Thief's Arm," in *Modern Street Ballads*, ed. John Ashton (London: Chatto & Windus, 1888).
142. Ibid., no lines.
143. "The Steam Arm," in "'The Steam Arm': Proto-Steampunk Themes in a Victorian Popular Song," by Kirstie Blair, *Neo-Victorian Studies* 3 (1) 2010: 204–205. In Blair's words, "'The Steam Arm' … is an anonymous ballad that appears to date from just before Victoria's accession to the throne. It relates the story, in the third person, of a one-armed Waterloo veteran, beaten by his 'arrant scold' of a wife, who determines to get even with her by replacing his lost arm with a mechanical steam-powered limb. The prosthetic arm is successfully grafted on to his stump, but once in action, it turns out to be too successful, not only striking down his wife, but the police who come to arrest him, the mayor who tries him, and the walls of the cell in which he is imprisoned. In the second-to-last stanza the hero attempts to return home and take his wife in his arms, but as his steam arm 'smash'd the crockery ware,' its incompatibility with love and domesticity is apparent. The soldier is finally doomed to remain alienated from all society, a haunting figure wandering 'like a sprite,' while 'his arm keeps moving with two-horse might.'" Kirstie Blair, "'The Steam Arm': Proto-Steampunk Themes in a Victorian Popular Song," *Neo-Victorian Studies* 3, no. 1 (2010): 196. Clear narrative and thematic overlaps with Hudson's "Cork Leg" and Parker's *The Flying Burgermaster* are apparent.
144. *I, Robot,* directed by Alex Proyas (2004; Walt Disney Studios HE, 2004), DVD.
145. Patricia Murphy, "The Fissure King: Parody, Ideology, and the Imperialist Narrative," in *The Victorian Comic Spirit: New Perspectives*, ed. Jennifer A. Wagner-Lawlor (Aldershot: Ashgate, 2000).

References

"The Advantages of Having False Teeth". *Judy*, Jan. 27, 1904.

Æsop, Allsopp. "The Wooden Leg and the Ungrateful Pensioner." *Funny Folks* 177 (1878): 126.

Akrich, Madeleine. "The De-Scription of Technical Objects." In *Shaping Technology/Building Society: Studies in Sociotechnical Change*, edited by Wiebe

E. Bijker and John Law, 205–24. Cambridge: Massachusetts Institute of Technology Press, 1992.

Anderson, George Kumler. *The Legend of the Wandering Jew*. Providence: Brown University Press, 1965.

Babbage, Charles. *An Essay on the General Principles Which Regulate the Application of Machinery to Manufactures and the Mechanical Arts*. London: W. Clowes, 1827.

Ballantyne, R. M. *Why I Did Not Become a Sailor*. In *Freak on the Fells; or Three Months' Rustication and Why I Did Not Become a Sailor*, by Robert Michael Ballantyne, 263–379. Philadelphia: The John Winston Co., 1864.

Bell, Henry G. "The Marvellous History of Mynheer Von Wodenblock." *The Edinburgh Literary Journal, or, Weekly Register of Criticism and Belles Lettres*, no. 47 (1829): 248–49.

———. *My Old Portfolio; or Tales and Sketches*. London: Smith, Elder, and Co., 1832.

Berkley, James. "Post-Human Mimesis and the Debunked Machine: Reading Environmental Appropriation in Poe's 'Maelzel's Chess-Player' and 'The Man That Was Used Up.'" *Comparative Literature Studies* 41, no. 3 (2004): 356–76.

Bigg, Henry Heather. *On Artificial Limbs, Their Construction and Application*. London: John Churchill, 1855.

Blackton, J. Stuart. *The Thieving Hand*. Veitagraph Company of America, 1908. YouTube. Accessed June 16, 2018. https://www.youtube.com/watch?v=J2Dr0MBWXgQ.

Blair, Kirstie. "'The Steam Arm': Proto-Steampunk Themes in a Victorian Popular Song." *Neo-Victorian Studies* 3, no. 1 (2010): 196–207.

Bly, Douglas. *A Description of a New, Curious, and Important Invention*. Rochester: Rochester Evening Express, 1859.

"Bolderoe's Widow". *All the Year Round* 17, no. 413 (1876): 157–68.

Boucicault, Dion. *After Dark: A Drama of London Life in 1868, in Four Acts*. New York: De Witt Publishing House, 1868.

Buxton, Harry Wilmot. *Memoir of the Life and Labours of the Late Charles Babbage Esq., F.R.S.* Cambridge: Massachusetts Institute of Technology Press, 1988.

Cooper, Arthur, dir. *Blind Man's Bluff*. 1903; Alpha Trading Company.

Corn, Joseph J. *User Unfriendly: Consumer Struggles with Personal Technologies, from Clocks and Sewing Machines to Cars and Computers*. Baltimore: John Hopkins University Press, 2011.

Cowan, Ruth Schwartz. "The Industrial Revolution in the Home: Household Technology and Social Change in the Twentieth Century." *Technology and Culture* 17 (1976): 1–23.

Crane, Frank. "Willie Westinghouse Invents an Automatic Arm." *The Wichita Daily Eagle*, April 10, 1904.

———. "Willie's 'Handshaker' Gets Papa into Trouble." *The San Francisco Call*, April 7, 1907.

"Wooden Leg as Weapon". *Daily Mail*, January 12, 1904.

Daly, Nicholas. *Literature, Technology, and Modernity, 1860–2000*. Cambridge: Cambridge University Press, 2004.

Daston, Lorraine J. "British Responses to Psycho-Physiology, 1860–1900." *Isis* 69, no. 2 (1978): 192–208.

Derrida, Jacques. "That Dangerous Supplement." In *Of Grammatology*, by Jacques Derrida, 141–65. Baltimore: John Hopkins University Press, 1997.

Dickens, Charles. *The Pickwick Papers*. New York: Hurd and Houghton, 1870.

———. *Dombey and Son*. London: Penguin, 2002.

———. *Our Mutual Friend*. London: Everyman, 2000.

———. "The Signal Man." *All the Year Round* 16, no. 400 (1866): 20–25.

Dickie, Simon. "Hilarity and Pitilessness in the Mid-Eighteenth Century: English Jestbook Humor." *Eighteenth-Century Studies* 37, no. 1 (2003): 1–22.

Dinello, Daniel. *Science Fiction Visions of Posthuman Technology*. Austin: University of Texas Press, 2005.

Doyle, Arthur Conan. *The Sign of Four*. Garches: Feedbooks, n.d.

Francis, M. E. "A Rustic Argus." *Longman's Magazine* 35, no. 210 (1900): 521–34.

Gavin, Adrienne E. "Dickens, Wegg, and Wooden Legs." *Our Mutual Friend*: The Scholarly Pages. Accessed June 18, 2018. http://omf.ucsc.edu/london-1865/victorian-city/wooden-legs.html.

Gerber, David A., ed. *Disabled Veterans in History*. Enlarged and revised ed. Ann Arbor: University of Michigan Press, 2012.

Girvan, Ray. "HG Bell: The First Cyborged Leg?" *JSBlog—Journal of a Southern Bookreader*. Last modified July 6, 2011. http://jsbookreader.blogspot.com/2011/07/hg-bell-first-cyborged-leg.html.

Green, Harvey. *Wood: Craft, Culture, History*. London: Penguin, 2007.

Hasegawa, Guy R. *Mending Broken Soldiers: The Union and Confederate Programs to Supply Artificial Limbs*. Carbondale: Southern Illinois University Press, 2012.

Hayes, Kevin J. "Visual Culture and the Word in Edgar Allan Poe's 'The Man of the Crowd.'" *Nineteenth-Century Literature* 56, no. 4 (2002): 445–65.

Helmholtz, Hermann von. "On the Conservation of Force." In *Nineteenth Century Science: A Selection of Original Texts*, edited by A. S Weber, translated by E. Atkinson *et al*, 280–99. Peterborough: Broadview, 2000.

Henham, Ernest G. "A Human Bundle." *Temple Bar* 111, no. 438 (1897): 42–58.

Herschbach, Lisa. "Prosthetic Reconstruction: Making the Industry, Re-Making the Body, Modelling the Nation." *History Workshop Journal* 44 (1997): 23–57.

Holmes, Oliver Wendell. "The Human Wheel, Its Spokes and Felloes." *Sounds from the Atlantic*, by Oliver Wendell Holmes, 282–327. Boston: Ticknor and Fields, 1864.

Hood, Thomas. *Miss Kilmansegg and her Precious Leg*. In *Selected Poems of Thomas Hood*, edited by John Clubbe, 193–270. Cambridge: Harvard University Press, 1970.

"How He Turned It with His Wooden Leg". *Illustrated Chips* 258 (1895): 4–5.
[Hudson, Thomas]. "The Cork Leg." [1819–1844] Broadside Ballads Online. Accessed June 21, 2018. http://ballads.bodleian.ox.ac.uk/view/edition/5413.
[Hudson, Thomas]. "Cork Leg." In *Modern Street Ballads*, edited by John Ashton, 153–55. London: Chatto & Windus, 1888.
Huxley, Thomas Henry. "On the Hypothesis That Animals Are Automata and Its History." *Nature* 10 (1874): 362–66.
Iguchi, Noboru, dir. *The Machine Gun Girl.* 2008; New York: Tokyo Shock, 2011. DVD.
Ketabgian, Tamara. *The Lives of Machines: The Industrial Imaginary in Victorian Literature and Culture.* Ann Arbor: University of Michigan Press, 2011.
Kipling, Rudyard. "Mrs. Bathurst." In Rudyard Kipling, *Traffics and Discoveries* (New York: Charles Scribner's Sons, 1909), 377–408.
"The Lame Landlord's Story". *Temple Bar* 20 (1867): 127–34.
La Mettrie, Julien Offray de. *L'Homme Machine.* Paris: F. Henry, 1865.
Levey, Jay, dir. *UHF.* 1989; Beverly Hills: MGM, 2002. DVD.
Lewis, Henry Clay. "The Indefatigable Bear Hunter." In *Louisiana Swamp Doctor: The Writings of Henry Clay Lewis*, edited by John Q. Anderson, 233–43. Baton Rouge: Louisiana State University Press, 1962.
Lindsay, Christina. "From the Shadows: Users as Designers, Producers, Marketers, Distributors, and Technical Support." In *How Users Matter: The Co-Construction of Users and Technology*, edited by Nelly Oudshoorn and Trevor Pinch, 29–50. London: Massachusetts Institute of Technology Press, 2003.
Lombroso, Cesare. *Criminal Man.* Edited and translated by Mary Gibson and Nicole Hahn Rafter. Durham: Duke University Press, 2006.
Marks, George E. *Marks' Patent Artificial Limbs with Rubber Hands and Feet.* New York: A. A. Marks, 1888.
Marten, James. "Nomads in Blue: Disabled Veterans and Alcohol at the National Home." In *Disabled Veterans in History*, edited by David A. Gerber, 324–47. Enlarged and revised ed. Ann Arbor: University of Michigan Press, 2012.
Marx, Karl. *Capital: A Critical Analysis of Capitalist Production.* Translated by Samuel Moore and Edward Aveling. London: Kimble & Bradford, 1938.
McKean, D. B. "A Wig and a Wooden Leg." *Chatterbox* 4 (1886): 27.
Melville, Herman. *Moby-Dick.* London: Penguin, 2012.
Menabrea, L. F. "Sketch of the Analytical Engine Invented by Charles Babbage." Translated with notes by Ada Lovelace. *Scientific Memoirs* 3 (1843): 666–731.
Mihm, Stephen. "'A Limb Which Shall Be Presentable in Polite Society': Prosthetic Technologies in the Nineteenth Century." In *Artificial Parts, Practical Lives: Modern Histories of Prosthetics*, edited by Katherine Ott, David Serlin, and Stephen Mihm, 282–99. New York: New York University Press, 2002.

Mitchell, David T., and Sharon L. Snyder. *Narrative Prosthesis: Disability and the Dependencies of Discourse*. Ann Arbor: University of Michigan Press, 2000.

Morris, Jenny. *Pride Against Prejudice: Transforming Attitudes to Disability*. London: The Women's Press, 1996.

Murphy, Patricia. "The Fissure King: Parody, Ideology, and the Imperialist Narrative." In *The Victorian Comic Spirit: New Perspectives*, edited by Jennifer A. Wagner-Lawlor, 23–42. Aldershot: Ashgate, 2000.

"Nineteenth- and Twentieth-Century Prostheses". The College of Optometrists. Accessed June 20, 2018. https://www.college-optometrists.org/the-college/museum/online-exhibitions/virtual-artificial-eyes-gallery/19th-and-20th-century-prostheses.html.

Norden, Martin F. *The Cinema of Isolation: A History of Physical Disability in the Movies*. New Brunswick: Rutgers University Press, 1994.

"New Use for a Wooden Leg". *Nottinghamshire Guardian*, June 17, 1899.

O'Connor, Erin. *Raw Material: Producing Pathology in Victorian Culture*. Durham: Duke University Press, 2000.

Otis, Laura. *Networking: Communicating with Bodies and Machines in the Nineteenth Century*. Ann Arbor: University of Michigan Press, 2001.

Ott, Katherine. "The Sum of Its Parts: An Introduction to Modern Prosthetics." In *Artificial Parts, Practical Lives: Modern Histories of Prosthetics*, edited by Katherine Ott, David Serlin, and Stephen Mihm, 1–44. New York: New York University Press, 2002.

———. "Hard Wear and Soft Tissue: Craft and Commerce in Artificial Eyes." In *Artificial Parts, Practical Lives: Modern Histories of Prosthetics*, edited by Katherine Ott, David Serlin, and Stephen Mihm, 147–70. New York: New York University Press, 2002.

Oudshoorn, Nelly, and Trevor Pinch, eds. *How Users Matter: The Co-Construction of Users and Technology*. London: Massachusetts Institute of Technology Press, 2003.

Packham, Catherine. *Eighteenth-Century Vitalism: Bodies, Culture, Politics*. Basingstoke: Palgrave Macmillan, 2012.

Park, Nick. *Wallace and Gromit: The Wrong Trousers. BBC*, 1993.

[Parker, Frances], the Countess of Morley. *The Flying Burgermaster: A Legend of the Black Forest*. n.p.: F. Morley, 1832.

"Peg-Legged Bob". *All the Year Round* 11, no. 280 (1874): 563–68.

Poe, Edgar Allan. "The Man That Was Used Up: A Tale of the Late Bugaboo and Kickapoo Campaign." In *The Works of the Late Edgar Allan Poe*, by Edgar Allan Poe, vol. 4, 315–25. New York: Blakeman & Mason, 1859.

Plunkett, John, Ana Parejo Vadillo, Regenia Gagnier, Angelique Richardson, Rick Rylance, and Paul Young, eds. *Victorian Literature: A Sourcebook*. Basingstoke: Palgrave Macmillan, 2011.

Proyas, Alex, dir. *I, Robot*. 2004; Walt Disney Studios HE, 2004. DVD.

Rabinbach, Anson. *The Human Motor: Energy, Fatigue, and the Origins of Modernity.* Berkeley: University of California Press, 1990.

Raimi, Sami, dir. *Evil Dead II.* 1987; Issy-les-Moulineaux: Studiocanal, 2019. DVD.

Rodriguez, Robert, dir. *Planet Terror.* 2007; Toronto: Entertainment One, 2008. DVD.

"A Romance of War". *Kind Words,* May 1, 1872: 145.

Rylance, Rick. *Victorian Psychology and British Culture: 1850–1880.* Oxford: Oxford University Press, 2000.

Shelley, Mary. *Frankenstein, or the Modern Prometheus.* London: Norton, 1996.

Shuttleworth, Sally, and Jenny Bourne Taylor, eds. *Embodied Selves: An Anthology of Psychological Texts 1830–1890.* Oxford: Oxford University Press, 1998.

Siebers, Tobin. "Disability as Masquerade." *Literature and Medicine* 23, no. 1 (2004): 1–22.

Smith, Roger. *Free Will and the Human Sciences in Britain, 1870–1910.* London: Pickering & Chatto, 2013.

Solnit, Rebecca. "The Annihilation of Time and Space." *New England Review* 24, no. 1 (2003): 5–19.

"The Steam Arm". In "'The Steam Arm': Proto-Steampunk Themes in a Victorian Popular Song," by Kirstie Blair, 204–205. *Neo-Victorian Studies* 3 (1) 2010: 196–207.

Sussman, Herbert. *Victorian Technology: Invention, Innovation, and the Rise of the Machine.* Santa Barbara: ABC-CLIO, 2009.

Swade, Doron. "'It will not slice a pineapple': Babbage, Miracles and Machines." In *Cultural Babbage: Technology, Time and Invention,* edited by Francis Spufford and Jenny Uglow, 34–52. London: Faber and Faber, 1997.

Sweet, Ryan. "'A Human Bundle': The Disaggregated Other at the *Fin de Siècle.*" *Victorian Review* 40, no. 1 (2014): 14–18.

———. "'Get the Best Article in the Market': Prostheses for Women in Nineteenth-Century Literature and Commerce." In *Rethinking Modern Histories of Prostheses in Anglo-American Commodity Cultures, 1850–1960,* edited by Claire L. Jones, 114–36. Manchester: Manchester University Press, 2017.

Thackeray, William Makepeace. "The Notch on the Ax—a Story a La Mode." *Roundabout Papers,* by William Makepeace Thackeray, 272–315. Sydney: ReadHowYouWant, 2009.

"The Thief's Arm". In *Modern Street Ballads,* edited by John Ashton, 150–52. London: Chatto & Windus, 1888.

Tonkelaar, I. den, H. E. Henkes, and G. K. Van Leersum. "Herman Snellen (1834–1908) and Müller's 'Reform-Auge': A Short History of the Artificial Eye." *History of Ophthalmology* 4 (1991): 349–54.

"Too Hard Upon My Aunt". *All the Year Round* 10, no. 242 (1863): 381–84.

Trollope, Anthony. *He Knew He Was Right.* Oxford: Oxford University Press, 2008.

Turner, David M., and Alun Withey. "Technologies of the Body: Polite Consumption and the Correction of Deformity in Eighteenth-Century England." *History* 99, no. 338 (2014): 775–96.

Twain, Mark. *Roughing It.* London: Penguin, 1988.

Uglow, Jenny. "Introduction: 'Possibility.'" In *Cultural Babbage: Technology, Time and Invention*, edited by Francis Spufford and Jenny Uglow, 1–23. London: Faber and Faber, 1997.

Ure, Andrew. *The Philosophy of Manufactures; or, an Exposition of the Scientific, Moral and Commercial Economy of the Factory System of Great Britain.* London: C. Knight, 1835.

"The Use of a Wig—a Hint to the Ladies". *The Satirist; or, the Censor of the Times* 675 (1845): 90.

Vaughn, Matthew, dir. *Kingsman: The Secret Service.* 2014; Beverly Hills: 20th Century Fox Home Entertainment, 2015. DVD.

The Walking Dead. AMC, 2010–.

"The Wandering Jew". *Chambers's Journal of Popular Literature, Science and Arts* 9, no. 458 (1862): 235–36.

Warne, Vanessa. "'If You Should Ever Want an Arm': Disability and Dependency in Edgar Allan Poe's 'The Man That Was Used Up.'" *Atenea* 25, no. 1 (2005): 95–105.

———. "'To Invest a Cripple with Peculiar Interest': Artificial Legs and Upper-Class Amputees at Mid-Century." *Victorian Review* 35, no. 2 (2009): 83–100.

Weber, A. S., ed. *Nineteenth-Century Science: A Selection of Original Texts.* Peterborough: Broadview, 2000.

Wells, H. G. *The Time Machine.* Oxford: Oxford University Press, 2017.

———. *The Island of Dr. Moreau.* London: Broadview, 2009.

———. *The Invisible Man.* London: Penguin, 2005.

———. *When the Sleeper Wakes.* Project Gutenberg. Last modified September 17, 2016. http://www.gutenberg.org/files/775/775-h/775-h.htm.

———. *The Food of the Gods and How It Came to Earth.* Project Gutenberg. Last modified August 14, 2010. http://www.gutenberg.org/cache/epub/11696/pg11696-images.html.

"The Wig Wouldn't Work like the Natural". *Illustrated Chips* 10, no. 248 (1895): 3.

Willis, Martin. *Mesmerists, Monsters, and Machines: Science Fiction and the Cultures of Science in the Nineteenth Century.* Kent: Kent State University Press, 2006.

"Wooden Legs". *The Court Magazine and Belle Assemblee* 2, no. 3 (1833): 147–51.

"A Wooden Leg That Knows a Thing or Two". *Every Week: A Journal of Entertaining Literature* 53, no. 1355 (1895): 21.

Woodforde, John. *The Strange Story of False Teeth*. London: Routledge and Kegan Paul, 1968.

Youngquist, Paul. *Monstrosities: Bodies and British Romanticism*. Minneapolis: Minnesota University Press, 2003.

Yuan, David D. "Disfigurement and Reconstruction in Oliver Wendell Holmes's 'The Human Wheel, Its Spokes and Felloes.'" In *The Body and Physical Difference: Discourses of Disability*, edited by David T. Mitchell and Sharon L. Snyder, 71–88. Ann Arbor: University of Michigan Press, 1997.

CHAPTER 4

Mobilities: Physical and Social

In the realm of social class, the norm is typically not the mean but the ideological fantasy of the mean. The fantasy is an ideological necessity if bourgeois capitalism is to project a positive vision of its operative world as free, prosperous, and coherent.

—Lennard J. Davis, *Bending Over Backwards: Disability, Dismodernism, and Other Difficult Positions* (New York: New York University Press, 2002), 96.

So the boy's artificial arm was taken away from him, and another of more approved and utilitarian pattern was given to him. This arm has a hand threaded at the wrist, so that it may be quickly unscrewed and taken off. In its place then appears every morning after the whistle blows a neat little hook, admirably adopted for engaging the handle of a water bucket. So now the one-armed boy is equipped for carrying two buckets of water instead of one, and the Carnegie Steel company has neatly adjusted what might have been a loss so that it begins already so that it figures of the credit side of the ledger.

—"The Mill Boy and the Hook." *The Spokane Press*, November 18, 1910, editorial page.

Lennard J. Davis is a leading name in cultural and literary disability studies whose work has been influenced by an upbringing by two parents who experienced deafness. In a personal essay titled "A Voyage Out (or Is It Back?)," published in his 2002 collection *Bending Over Backwards*, Davis

R. Sweet, *Prosthetic Body Parts in Nineteenth-Century Literature and Culture*, Palgrave Studies in Nineteenth-Century Writing and Culture, https://doi.org/10.1007/978-3-030-78589-5_4

notes how his parents' lives made him attentive to the ties between disability and social mobility: "To me, my parents' deafness will always be inseparable from our social class."[1] Here, Davis reminds us that the constructed notion of the normal body is ideologically tied to the modern economic model. The link between high social status and the normal body is evidenced to Davis, in part, by personal experience. He explains that in the 1950s, when he was growing up, people with deafness were often factory workers, his father being one of many in such a role.[2] The link between disability, the factory (a technology that was conceptualized in the nineteenth century as a kind of the prosthesis), and lower social status is exemplified by the 1910 sketch "The Mill Boy and the Hook." This piece exposes both the restrictions imposed by capitalist society on the disabled body and the extent to which the prosthesis, both conceptually and materially, has often been tied to commercial purposes.[3] The hook hand in this tale is a material solution to a physical problem implemented not so much for the "one-armed boy" as for the economic benefit of the factory owner; if "the hand" served—as in Charles Dickens's *Hard Times* (1854)—as synecdoche for the factory worker, whose labour is concentrated in a useful appendage, then here is an amplification of the worker-as-tool logic.[4] This chapter addresses literary sources that illuminate the trifold relationship between disability, technology, and social class.

The social model of disability has shown us how stigma, exclusion, and the ableist construction of our built environment have restricted opportunities for impaired people in social terms, but limited work has been done that explores the class trajectories of disabled figures in imaginative works. What are the imagined fates of impaired characters who try to improve their social position? Do prostheses enable social mobility, as real-life makers promised, or do they restrict the user socially? What do fictional representations tell us about the position of prosthesis users in class hierarchies?

At a time when prosthetic devices began to saturate the marketplace and occupy a greater place in the social consciousness than ever before, public opinion regarding the respectability of artificial body parts remained largely undecided. Stephen Mihm has shown us that limbs "which shall be presentable in polite society" were required for those in the upper classes who lost limbs and wished to maintain social distinction.[5] Similarly, Erin O'Connor suggests that artificial limbs functioned for workers as material solutions to the rupture in the physical economy supposedly occasioned by limb loss.[6] Still, cultural and literary representations, relatively overlooked aspects of the discursive history of prostheses, reveal that criticism was also levied towards artificial body parts—technologies that some saw as "emblem[s] of deceit."[7] Certain primitive devices, such as peg legs, became regularly associated with beggary. Meanwhile, fictional works, including Charles Dickens's *Our Mutual Friend* (1864–1865), explore the extent to which a fusion with technology, as manifest by the wooden-leg user Silas Wegg, could enable one to become upwardly mobile.[8]

By turning to Dickens's novel as a case study alongside several related but now lesser-known contemporary writings, this chapter shows how nineteenth-century fiction complicates the emerging hegemony of physical wholeness by revealing certain paradoxes and imagining upwardly mobile characters who defied conventions associated with bodily normalcy. The privileging of physical completeness that produced a demand for lifelike prosthetics and linked the normal body to social success was problematized by fictional works that imagined conspicuously "crippled" wooden-leg-using characters as ironic exemplars of social mobility. Although many of the sources discussed come to conservative conclusions that reify normative bodily standards, the ideological necessity for such a rebuttal paradoxically reveals the conceptual fragility of physical completeness as the hallmark of the healthy body.

Mendicity Versus Mendacity

One critical context for the relationship between prosthesis use and class mobility is the nineteenth-century physical economy of work. As O'Connor explains:

> Victorian ideals of health, particularly of male health, centered on the concept of physical wholeness: a strong, vigorous body was a primary signifier of manliness, at once testifying to the existence of a correspondingly strong spirit and providing that spirit with a vital means of material expression. Dismemberment disrupted this physical economy.[9]

In a world in which not only normative bodily function but also an impression of physical wholeness was so heavily valued, those who were missing limbs often found work hard to come by. Cindy LaCom has shown how, among other factors, "public perceptions of and responses to people with disabilities and to the very concept of disability were shaped by … developing capitalist economic theories and an ideology of self-help," and "the growth of industrialism."[10] As she explains: "Those unable to meet industrial workplace standards because of a disability or deformity were increasingly exiled from the capitalist 'norm', which demanded 'useful' bodies, able to perform predictable and repeated movements."[11] To borrow a quotation that LaCom uses, in 1846 a factory inspector noted that "sound limbs are a main part of the working man's capital, and they should be exposed as little as possible to the risk of irrevocable diminution."[12] In a world in which an individual's productivity was seen as an index of character, convincing those around you that you could work hard was important. As Marta Russell observes: "disabled people who were perceived to be of no use to the competitive profit cycle would be excluded from work."[13] For industrial labourers, such as railway or factory workers, it was important to continue to work after injury if possible. This point has been made by Jamie L. Bronstein: "For workers, injury lasted as long as it kept one from returning to the same or a similar job at the same pay rate."[14]

Nonetheless, many positions, especially higher-paid ones, were considered untenable for those who had lost arms or legs. For instance, railway owners tended to relegate disabled workers to lower-paying positions outside of public view to avoid offending public sensibilities.[15] Writing for the *Miners' Journal and Pottsville General Advertiser* in 1846, one disabled railway worker wrote, "A man whose frame is shattered to such an extent as to render him a cripple for the remainder of his existence, is practically dead so far as active work is concerned."[16] Likewise, as Mihm remarks, "[i]n an age of appearances, members of the middle-class necessarily hid their deformities and weaknesses, for fear that first impressions might deny them opportunities in marriage, employment, and social advancement."[17] We learn from French mid-century artificial-eye maker Auguste Boissonneau that prejudices were not restricted to those who had lost limbs. According to him, the use of one of his devices "permit[ted] its wearer to look after his business and keep up his relations with society in general, without the fear of being looked upon as an object of repulsion or of pity."[18]

The social situation for those who lost body parts but could continue to work was restrictive, but for those who were completely excluded from the working environment by perceived bodily loss, the possibility of social improvement was often erased entirely. As Martha Stoddard Holmes explains, both state and charity relief for those deemed unemployable due to disability were jeopardized by anxieties surrounding who it was that *deserved* financial assistance.[19] The discourse of disability and financial provision was often infiltrated by fears of fraudulence. In 1834, the Poor Law Amendment Act introduced a centralized system to manage administering relief to the poor. The new system was brought about, at least in part, in response to concerns from social reformers that the parish-based system of poor relief implemented by the Elizabethan poor laws was inconsistent and, at times, misused. The new system sent able-bodied men seeking aid to the workhouse while providing limited out relief to those deemed deserving—those unable to work, including the young, the old, and the disabled. Despite this tightening of the law, as Holmes reveals, fears of abuse remained and much debate surrounding the deservingness of the unwaged disabled perpetuated as a result.[20] Even charitable discourse was pervaded by these questions. For instance, the Charity Organisation Society was set up in 1869, inspired by the belief that "the mass-misery of great cities arose mainly, if not entirely, from spasmodic, indiscriminate, and unconditional doles, whether in the form of alms or in that of Poor Law relief."[21] This situation meant that many unwaged disabled people were not awarded relief for fears of malingering, thereby prohibiting their chances of bettering their social position. As Holmes comments: "even if a physical impairment looks valid, the argument goes, it may be supplemented by an invisible advantage: one more instance of the difficulty of distinguishing mendicity from mendacity."[22]

Fears that homeless people feigned disability to secure alms or cheat the poor relief system were buttressed by a culture that perpetuated images of fraudulent beggars. These representations remained visible throughout the century in England and elsewhere in the Western world. Susan M. Schweik notes, for instance, how the duplicitous mendicant was a common image in mid-to-late nineteenth-century American culture: "The distinction between false and true mattered enough to produce a tension between languages of care and languages of criminality, and conflict about authenticity, that played themselves out over and over in the telling of stories of the vagrant and the beggar."[23] Meanwhile, in England Henry Mayhew's expansive and popular exercise in social observation *London Labour and the London Poor*—a work that taxonomized London's labourers, street workers, and non-workers—identified a class of mendicants as "Those that will not work."[24] In categorizing and describing the kind of beggars that inhabit the streets of London, Mayhew discusses "Blown-up miners" as a group rife with impostors. He describes how some "rank impostors" expose "some part of their bodies—the leg or the arm—and show you what looks like a huge scald or burn." Moreover, he labels the device of producing artificial sores "scaldrum dodge."[25] Beggars feigning disability also appear in Victor Hugo's grotesque masterpiece *Notre-Dame de Paris* (1831).[26] Drawing from this trope, Arthur Conan Doyle's Sherlock Holmes story "The Man with the Twisted Lip" (1891) recounts the scandalous career of Neville St. Clair, a well-to-do gentleman who, with the aid of tape, which he uses to make his mouth look disfigured, turns secretly to begging: a pursuit that he finds easier and more profitable than other professions available to him.[27] While the image of the begging impostor was by no means new in the nineteenth century, the popular motif of the fraudulent mendicant in Victorian literature and culture paid testament to the prevalence of cultural misgivings about aberrant bodies in this period.[28]

In addition to cultivating suspicions about the authenticity of the disabled poor, journalistic works, such as Mayhew's *London Labour and London Poor*, gave disabled individuals a chance to share their life experiences. A section of Mayhew's first volume tells the story of a "crippled" seller of nutmeg graters, a man with deformed limbs who is unable to walk and perform other basic motor skills.[29] We learn through this example of the social limitations and difficulties imposed upon an individual whose body is unaccommodated for by society. In a lengthy personal testimony, the seller of nutmeg graters reveals how he ended up working as a street trader. We learn that various circumstances, many linked to disablism, led to his downfall: his "feeble-minded" mother was unable to look after him, entrusting a colleague (a fellow servant) to be his guardian. His extended family were repulsed by him and thus unwilling to provide financial assistance. His former lodgers exploited him by refusing to pay rent, leading to his financial collapse and stay in a workhouse. In the workhouse, he was unable to perform manual labour, resulting in his food allowance being restricted, thereby worsening his physical condition and prohibiting him from saving enough capital to buy sufficient stock for costermongering. The seller of nutmeg graters also explains how physically taxing his work is:

> It's very hard work indeed is street-selling for such as me. I can't walk no distance. I suffer a great deal of pains in my back and knees. Sometimes I go in a barrow, when I'm travelling any great way. When I go only a short way I crawl along on my knees and toes. The most I've ever crawled is two miles. When I get home afterwards, I'm in great pain. My knees swell dreadfully, and they're all covered with blisters, and my toes ache awful. I've corns all on top of them.[30]

Due to the unprofitable nature of his work ("Some weeks [he] hardly clear[ed] [his] expenses"), lack of support network, unwillingness to beg for alms, and inability to secure other employment, it appears inevitable that this individual will spend the remainder of his days on or below the breadline.[31] Earlier "autobiographical" narratives by "factory cripples," such as Robert Blincoe and William Dodd, add further insight regarding the social prospects of disabled men from the lower classes.[32] Blincoe's and Dodd's respective narratives describe how public perception of their impairments provided barriers to work. Blincoe reveals how he was paid less than nondisabled colleagues while Dodd notes how he was refused work as an errand boy to an ironmonger and later as a teacher due to prejudices attached to his physical difference. After the amputation of his lower arm, Dodd resigned himself to a life "unfit for … business."[33] These accounts provide insights into the potential social plight facing those who experienced impairment in the nineteenth century.

Despite the heart-rending nature of these stories and Mayhew's apparent admiration for the resilience and industriousness of the seller of nutmeg graters, as in many representations of disabled street dwellers from this era, fears of malingering remained on the surface. Mayhew notes that he made "all due inquiries" to satisfy himself of the seller of nutmeg grater's "worthiness."[34] Moreover, Mayhew provides several testimonies after the street seller's personal account to convince himself and the reader of the man's honesty. This latter measure exposes the lack of trust that Mayhew placed in those who display physical difference in Victorian London, revealing how questions of authenticity often pervade Victorian discussions of the disabled adult male body.[35]

Intriguingly, the very notion that one could better their social position by imitating disability, discloses how nineteenth-century discourses of disability and prosthesis use brought into doubt the dominance of the nondisabled, physically complete body. Here, the so-called able-bodied are concerned with and resistant to a method of self-representation that privilege physical loss over completeness for monetary ends. Thus, while physical wholeness is so often privileged in Victorian discourse, it emerges as an unstable conceptual category, in part because so many deviated from it in reality—hence its need for repeated cultural reinforcement.

Both questions of authenticity and the social trajectories of people with disabilities were topics central to nineteenth-century discourses of prosthetics. Relating to the latter, prosthetic limbs were often marketed as devices that could enable users to avoid the social plight that the seller of nutmeg graters describes. Still, not all devices were affordable for those who needed them the most. The next section explores the distinction between peg legs and artificial legs, lower-limb prostheses available, for the most part, to consumers of distinct socio-economic groups.

The Peg Versus the Artificial Leg

The promise of social security was used as part of a commercial advertising rhetoric employed by contemporary prosthetists. As O' Connor contends: "The discourse of prosthesis is ... infused with class consciousness, suggesting that man cannot occupy a meaningful social position unless he is physically complete."[36] Prosthesis makers often cited the social disadvantages of physical difference in contrast to the mimetic capacities of their devices, which they claimed could mask the appearance of impairment (thus alleviating stigma and social degradation) and allow the user to occupy a social position usually reserved for the normate. French ocular prosthesis maker Boissonneau states in a treatise on artificial eyes that his prosthesis design, "whilst concealing a deformity of, to say the least, a disagreeable aspect, permits its wearer to look after his business and keep up his relations with society in general, without the fear of being looked upon as an object of repulsion or of pity."[37] Elsewhere, the maker goes even further, suggesting that the use of one of his prosthetic eyes could enable its wearer to make a more favourable impression upon those whom s/he encountered: "The use of an artificial eye is highly appreciated by those who know how much the facial defects are in the way of one's progress in the world, and how painful is the contention between the unpleasant impression caused by an unbecoming face, and the wish of pleasing which every one experiences."[38] Along similar lines, in *Automatic Mechanism*, the limb prosthetist Frederick Gray refers to the fitting of an artificial limb as "the facility of progression," once again implying that those missing body parts needed prostheses to move forward in life—in terms of both physical locomotion and social mobility.[39] In his treatise, Gray also includes a great number of testimonial letters, which in many cases emphasized how his artificial limbs enabled users to maintain their social position. Captain W. W. A. from the Bengal Army, for instance, states: "To one in my situation, the benefit I have derived from your skill is incalculable, as it enables me to return to an efficient performance of my military duties."[40]

Making life for the amputee even more problematic, the type of prosthesis employed by an individual was seen as a measure of social standing. Vanessa Warne helpfully explains the distinction between peg legs and their more sophisticated counterparts:

> The term "artificial leg" was reserved for prostheses that imitated both the appearance and movement of a natural leg; it did not apply to simple wooden pegs or to rudimentary leg-shaped prostheses. Marketed as more attractive, comfortable, and safe than crutches or pegs, artificial legs had patented features such as rubberized feet and articulated joints. They were usually made to order and were consequently costly.[41]

For prosthetists, such as Gray, it was the specialist skill of artificial-limb makers that differentiated their devices from rudimentary pegs: "the science of artificial limb-making is neither a very simple nor easy acquirement; it cannot be attained without great attention, great experience, and a habit of induction applied to facts."[42] American poet and physician Oliver Wendell Holmes suggests it is the devices made by these skilled professions that are deemed acceptable "in polite society."[43] Holmes famously writes in "The Human Wheel, Its Spokes and Felloes" (1864) that while "[a] plain working-man, who has outlived his courting-days and need not sacrifice much, to personal appearance, may find an honest, old fashioned wooden leg, cheap, lasting, requiring no repairs, the best thing for his purpose," in "higher social positions at … an age when appearances are realities … it becomes important to provide the cripple with a limb which shall be presentable in polite society, where misfortunes of a certain obtrusiveness may be pitied, but are never tolerated under the chandeliers."[44] As Holmes observes, while the old-fashioned peg served as an acceptable device for lower-class users to wear, its crude construction meant that it was utterly unsuitable for the respectable amputee for whom it was a social requirement to display good health—a condition that relied upon an impression of physical wholeness. What Holmes fails to acknowledge is that though peg legs might have been aesthetically acceptable within a working-class environment, their lack of functionality rendered them extremely limited devices. Peg-leg users were often excluded from manual jobs because of assumptions made about their limited physical abilities. Furthermore, the general unaccommodating design risked further impairing amputees.

It comes as no surprise, therefore, that peg legs were the cheapest and most basic devices available to those who had lost legs. Having been in circulation for several centuries, these devices consisted commonly of hollow wooden buckets or cups into which the stump of the user was placed. The support was fixed to a wooden peg that would reach the ground instead of a foot. As an article published in *All the Year Round* in 1875 comments: "In order to make it look a little more shapely than a mere stick of firewood, the peg is contoured somewhat in rolling-pin fashion, with a knob at the lower end."[45] For slightly more money, other adaptations could also be made to better fit the amputee's stump: a leather sheath could be attached for improved comfort or a knee joint could be incorporated so that the peg could be bent whilst sitting-down. Despite these possible modifications, however, the peg-leg was considered a low-end product: "what are wooden-pegs compared with artificial legs? No more than penny dolls compared with Mr Cremer's walking and talking young ladies."[46] Since these devices were most often used by working-class or pauper amputees, they became synonymous with low status and small income. As if to suggest that the beggar with wooden leg was a ubiquitous figure, one *All the Year Round* contributor writes: "You, London reader, have seen wonderful things in your time; the sham sailor in the New-road, with a painting of a storm in the Bay of Biscay rolled out between his wooden legs, which rest as sentinels on either side of it."[47] As we will see later in this chapter, this reporter's apparent enthusiasm for potentially fraudulent wooden-leg users of the streets extended from Dickens's journalism to the pages of his novels.

With regard to the class-related bifurcation of lower-limb prostheses, a detail that has thus far been overlooked by scholars of Victorian prosthesis is that, while great advances were made in artificial limbs over the course of the nineteenth century, the vast majority of cultural and fictional representations of lower-limb prostheses describe peg legs rather than the costly, new-fangled devices made by emerging prosthetists.[48] In this sense, then, cultural and literary representations of leg prostheses convey the real-life paradox in nineteenth-century prosthesis circulation. In a society that demanded working men to look productive and thus whole, the consumer group that required the use of effective functional prostheses the most was the labouring class, who were often injured by industrial machinery or who lost limbs to disease or infection. In a study of the dangers of industrial work to the labourer's body, Jamie L. Bronstein reveals: "Construction work on a single stretch of railway line over a six-year period resulted in 32 deaths, 23 compound fractures, 74 simple fractures, and 140 'severe cases', including blast burns, severe bruises, cuts, and dislocations. Many of those injured suffered from multiple injuries." These injuries were "in addition to 400 cases of minor accidents: trapped and broken fingers, seven of which had to be amputated; injuries to the feet, lacerations of the scalp, bruises and broken shins."[49] Furthermore, by providing an indication of the high percentage of amputations performed on those from working-class positions, E. J. Chaloner, H. S. Flora, and R. J. Ham show us that of the eighty-four amputations performed at the London Hospital between 1852 and 1857 as a consequence of trauma, "72 were necessitated by injuries sustained at work—for instance, 'being run over by a railway car', being 'crushed between two ships' or being 'injured by machinery.'" These authors note that "[t]he distribution of injury is reflected in the occupations of the male patients, most of whom were labourers, railwaymen, sailors or factory workers."[50] Providing a personal and geographical perspective, in 1841 the self-named "factory cripple" Dodd recalled that "[a]ccidents by machinery in the North are of a week, nay, almost daily occurrence."[51] As I have already noted, it was a financial imperative for injured labouring-class individuals to return to manual work as soon as possible. Despite their apparent need to appear able-bodied in order to convince supervisors of their ability to perform, these amputees simply could not afford the kinds of devices that would provide them with the best chance of passing as normal.

Prosthetists were fully aware of the large consumer group that injured workers presented, and some commentators, such as Gray, were frustrated that labourers' limited means excluded them from the artificial-limb market. Gray, for instance, laments that unless they were sponsored by a particularly generous employer, "from the expense entailed by their elaborate construction, [artificial limbs] are not within the reach of the poorer class of sufferers." It is with regret that Gray makes this declaration since, as he observes, "in the case of the affluent the loss of a limb does not reduce the sufferer to a state in which his relative position in life is rendered worse, whereas, when a poor man becomes crippled, he is reduced to a state of almost perfect destitution and misery."[52] For Gray, then, amputation is more debilitating for a labourer than for a member of the aristocracy or middle class, primarily because persons belonging to these groups did not need to rely so heavily on their physical capacities to earn an income. Despite Gray's encouragement for employers to fit their injured workers with more suitable prostheses—no doubt at least in part inspired by commercial interests—in the majority of cases, working-class amputees were forced to make-do with peg legs.[53] Underlining the way in which some prosthetists prioritized profit-making over care for working-class clients, in their 1888 catalogue, the internationally successful American firm A. A. Marks renounced the use of peg legs before a few sentences later stating that they could produce peg legs upon request.[54] We learn from statements including this that working-class amputees were a prevalent but not primary consumer group for the burgeoning prosthesis market due to their financial limitations.

Unlike the artificial leg, which was designed so that its non-human composition could be easily concealed, the peg leg was obviously wooden.[55] It is therefore worth taking a moment to consider the contemporary connotations of this material, since woodenness was often so lucidly displayed by peg-leg users. While we may be inclined to assume that the industrial revolution rendered wood archaic, as Harvey Green reveals, even by the mid-nineteenth century, timber remained prominent:

> we should remember that the Industrial Revolution in the West began with wood as its major material. Before 1850 most machines—spinning wheels, looms, plows, rakes, shovels, hoes, churns—were made almost entirely of wood. Even at the outset of the age of steel, machines for home and factory production were made mostly of wood, with iron or steel fittings attached at areas of greatest friction or where cutting took place. But iron and steel require fuel (wood, charcoal, coal, coke) to smelt the ores and melt the metals. Wood requires no further transformation of its substance in the wild. Metals and plastics may be the materials of industrialism today, but wood made the revolution possible.[56]

An article published in the *London Journal* in 1862 pays testament to the enduring uses of wood, not the least as a means of fuel and as a durable building material.[57] Thus, while today wood "endures because it is now thought of as a traditional, even (ironically) preindustrial material," in mid-Victorian times it was contiguous with industrial practice.[58] Chiming with David Edgerton's innovative research on "old" technology in use, wood was far more entwined with industrial modernity than often accredited.[59] To many Victorians, the wooden-leg user therefore appeared not only as someone who was not fully human in composition, but also as a person whose body part was representative of industrial manufacture—made by and appearing a part of an increasingly autonomous mechanical system. As we will see in the following sections, links with industry complicated representations of wooden-leg users, who were sometimes imagined as transgressive social climbers.

Prosthesis Users in Dickens's Journals

It can be confidently asserted that Dickens was well-accustomed to editing and imaginatively creating works including prosthesis-using characters well before the first instalment of *Our Mutual Friend* was published in May 1864. As Adrienne E. Gavin asserts: "while his novels are not filled 'entirely by wooden legs,' the number of wooden legs within them reveal his fascination with these limbs."[60] Discussing Dickens's deployment of effigy and his interest in the boundary between human and thing, John Carey remarks that "Dickens' most popular lifeless bit is the wooden leg, about which he has a positive obsession."[61] Furthermore, in *Dombey and Son* (1846–1848) the author presents to the reader the hook-hand user Captain Cuttle, a character who, according to Herbert Sussman and Gerhard Joseph, "literalizes Dickens's sense of the emerging prosthetic man."[62] Wooden legs are also referred to or appear in *The Pickwick Papers* (1836–1837), *Nicholas Nickleby* (1837–1838), *The Old Curiosity Shop* (1840–1841), *Barnaby Rudge* (1841), *Martin Chuzzlewit* (1842–1844), and *David Copperfield* (1849–1850).[63]

In *Nicholas Nickleby*, Miss Knag tells Madame Mantalini about her uncle, who "had such small feet, that they were no bigger than those which are usually joined to wooden legs."[64] *The Pickwick Papers* depicts an amputee who is a reformed alcoholic. He is reported to have found "a wooden leg expensive going over the stones," so for a while wears second-hand legs and drinks "a hot glass of gin and water every night—sometimes two." After finding that "second-hand wooden legs split and rot very quickly," he is "firmly persuaded that their constitution was undermined by the gin and water." He quits drinking and buys new wooden legs, which he finds "last twice as long."[65] In *The Old Curiosity Shop*, freak-show proprietor Mr. Vuffin asserts, "Look at wooden legs. If there was only one man with a wooden leg what a property he'd be! … Instead of which … if you was to advertise Shakespeare played entirely by wooden legs, it's my belief you wouldn't draw a sixpence."[66] Simon Tappertit, from *Barnaby Rudge*, is crushed in a riot thus leaving him with two wooden legs.[67] Mrs. Gamp, from *Martin Chuzzlewit*, describes how her husband would send his son on an "errand to sell his wooden leg for any money it would fetch as matches in the rough, and bring it home in liquor."[68] *David Copperfield* also features a wooden-leg user: Mr. Tungay, the "obstinate barbarian" henchman of the violent schoolmaster Mr. Creakle.[69] The tale also includes the miser Mr. Barkis, who is found to own "a silver tobacco-stopper, in the form of a leg," perhaps drawing from Thomas Hood's popular poem *Miss Kilmansegg and Her Precious Leg* (1840–1841).[70]

Dickens goes a step further in his interrogation of human-machine boundaries in a section entitled "Display of Models and Mechanical Science" of *The Mudfog Papers,* which was published in *Bentley's Miscellany* from 1837 to 1838. Here, in a satirical dig at the British Association for the Advancement of Science, the ironically named Mr. Coppernose, a member of the "Mudfog Association for the Advancement of Everything," proposes creating an automaton police force for the relief of carousing young noblemen who were inclined to "pummelling each other."[71] Jay Clayton draws attention to the author's bizarre obsession with man-made bodies and body parts with his assertion that "Dickens seems to find something grotesque in the very idea of automata."[72] Dickens's interest in prostheses is corroborated by his correspondence. In a letter to John Leech from 23 October 1848, Dickens writes of his enthusiasm to see "a gentleman with a wooden leg … dance the Highland Fling" as advertised in a Britannia Saloon Bill.[73]

Several prosthesis-using characters also appeared in *Household Words* (1850–1859), a weekly magazine established and edited by the author. These representations include the dust scavenger with wooden leg, Peg Dotting, from Richard H. Horne's "Dust; or Ugliness Redeemed" (1850); a wooden-leg user described in Dickens's article "New Year's Day" (1859); and the respectable amputee milliner's daughter Mary Wigley from Sarah Smith's short story "The Lucky Leg" (1859)—to mention but a few.[74] William Blanchard Jerrold's famous article on the Victorian penchant for prostheses "Eyes Made to Order" also appeared in *Household Words* in 1851. What these examples demonstrate is not only that Dickens was familiar with cultural and fictional representations of lower-limb amputees but that he was also preoccupied with issues relating to prosthesis use.[75] These imaginaries explored the increasingly intimate relationship between people and things, which Dickens was keen to scrutinize. As I show, various works that Dickens wrote or was closely associated with complicated the cultural dominance of the physically whole body by portraying socially mobile prosthesis users.

Horne's "Dust; or Ugliness Redeemed" is a tale that is interesting not the least because it is thematically and narratively similar to *Our Mutual Friend*. Like Dickens's later work, Horne's narrator associates a wooden-leg user with dust—the detritus of Victorian London that Dickens used to explore man's bizarre yet increasingly intimate relationship with the material world. In "Dust," three physically non-typical members of the underclass, with the help of a supposedly enchanted dust heap, work together to rescue and revive a social superior who has fallen into a nearby canal. The first of this story's unconventional heroes is the ninety-seven-year-old Gaffer Doubleyear, who wears an oyster shell over a missing eye and whose name is partially recycled for the like-minded, though more sinister, scavenger Gaffer Hexam in *Our Mutual Friend*. Doubleyear is accompanied by the aptly named Peg Dotting, an eighty-three-year-old wooden-leg user, and Jem, a "poor deformed lad whose back had been broken as a child."[76] In a stroke of good fortune, after being revived, the rescued gentleman notices that a piece of parchment wrapped around a treasure recovered by Jem is in fact a missing part of a title-deed. Anticipating the significance of the dust heap in *Our Mutual Friend*, this discovery provides a timely lift to the gentleman, who we learn has been through a period of hardship up to this point. Because of this discovery, he is able to recover socially and regain his former position. In gratitude to Doubleyear and Dotting, the gentleman purchases them a cottage. As we will see, like Wegg, these prosthesis-using characters experience an elevation in status thanks to their intimate relationship with the material world (for which they are marked physically by their use of prostheses).

In complicating the hegemony of natural physical wholeness, the idea that class mobility could be enabled as result of a close human-material relationship is explored and interrogated by Dickens's rewriting of the dust-heap story in *Our Mutual Friend*. However, unlike the prosthesis-using duo of "Dust"—who receive a moderate elevation in status as a result of their dust-heap knowledge and tender human generosity (though they become home owners, they continue to work as "scavengers" on the dust heap)—Dickens's novel inverts this narrative: *Our Mutual Friend* suggests that those who become too closely enmeshed with material objects ultimately fall because their human qualities decline. *Our Mutual Friend* brings into question the primacy of organic wholeness by imaginatively exploring class mobility enabled by prosthesis use while manifesting a distrust towards too close an intimacy with technology. In ways that remind us of Karl Marx's conceptualization of the human "organs" within the factory system, Wegg's humanity is compromised by the hardening influence of a technological device.[77]

Our Mutual Friend

Our Mutual Friend, first published in serial form from 1864 to 1865, is a novel that literalizes George Henry Lewes's claim that Dickens's characters are wooden puppets brought to life by incident.[78] The claim is that Dickens uses Silas Wegg, a man who is both a peg-leg user and depicted as wooden in appearance, character, and action, to explore the greed-provoking influence of material culture and to examine the privileging of organic physical normalcy.[79] Dorothy Van Ghent observes that Dickens's writing responded to a culture in which "[p]eople were being de-animated, robbed of their souls, and things were usurping the prerogatives of animate creatures."[80] *Our Mutual Friend* is clearly a novel that fits this description, since it displays an explicit engagement with topics relating to human/object relationships. Significantly, the novel imaginatively engages with prostheses, their materiality, and how they might impact social mobility.

Dickens's portrayal of Wegg complicates the normative physical models presented by contemporary influential figures such as Samuel Smiles, John Ruskin, and Charles Kingsley. These figures believed that displaying good physical health was a signifier of mental vigour, whereas Dickens imagined a conspicuous amputee who, for a time, is enabled class mobility thanks to his use of a wooden leg. In his highly successful *Self Help* (1859), Smiles asserts that the male body, alongside masculine morality and intellect, must be cherished and refined: "Each must be developed, and yet each must yield something to satisfy the claims of the others .... It is only by wisely training all three together that the *complete* man can be formed" (emphasis added).[81] For Smiles, physical cultivation is a vital yet overlooked aspect of complete and thus ideal masculinity. Smiles later asserts that "[i]t is in the physical man that the moral as well as the intellectual man lies hid; and it is through the bodily organs that the soul itself works."[82] Moreover, Smiles claims that the success of professional men "depends in no slight degree on their original stamina and cultivated physical strength."[83] Smiles's vision was reaffirmed by Ruskin, who in 1860 asserted that a gentleman's status derives from his physical vigour: "A gentleman's first characteristic is that fineness of structure in the body, which renders it capable of the most delicate sensation."[84]

Around the same time that Smiles and Ruskin promulgated their ableist masculine ideals, Kingsley re-envisioned masculinity infamously with his concept of "muscular Christianity." "Perhaps more than any other middle-class writer," James Eli Adams has suggested, "Kingsley placed the male body into widespread circulation as an object of celebration and desire."[85] Donald E. Hall claims that the defining characteristics of "muscular Christianity" included "an association between physical strength, religious certainty, and the ability to shape and control the world around oneself."[86] By emphasizing the importance of physical cultivation in his essay "The Science of Health" (1872), Kingsley proclaims: "We must … have the CORPUS SANEM if we want the MENTEM SANEM; and healthy bodies are the only trustworthy organs for healthy minds."[87]

Wegg would not be considered physically fit by Victorian standards and thus neither vigorous nor resolute. Scenes featuring Wegg such as the convulsive-stump episode, which I have explored in Chap. 2, place the amputee far from the mid-century physical ideal. However, in an entirely atypical manner Wegg's hardiness and desire for upward mobility are affected by stubbornness and inhumanity, qualities that are shown to stem not from the character's physical robustness, but instead from his use of a wooden leg. Wegg's initial rise is apparently enabled by his prosthesis and the supposed influence that the materiality of the device asserts.

Engaging with contemporary debates surrounding the relationship between man and machine in this period, while harnessing the oft-exploited potentially of physical difference for metaphorical power, Dickens presents Wegg as someone who, in almost every respect, resembles the wood that forms his prosthetic limb. It is, in part, the very fact that Wegg wears a wooden leg that convinces Mr. Boffin to employ him, thus enabling the amputee to improve his social status. However, though the author allows the wooden Wegg a great deal of social mobility prior to the novel's denouement, Dickens eventually privileges the organic over the artificial: Wegg's plan to usurp Mr. Boffin from his elevated position eventually fails—in part due to the inability of the amputee's peg leg to successfully negotiate the terrain of his master's dust mounds. Echoing the contemporary instructions that encouraged amputees to utilize sophisticated prosthetic devices to avoid the stigma surrounding bodily loss, Wegg's leg ultimately stands, metonymically, for stasis, reflecting the real-life social limitations imposed on wooden-leg users. Since Wegg's wooden prosthesis has such a profound impact on its wearer, Dickens's portrayal suggests both an uneasiness regarding the meshing of the human with the non-human and a lack of sympathy towards pauper amputees, who were so often depicted as figures of fraudulence and imposture in contemporary culture.

Prior to the uptake of disability theory within Victorian studies, several critics commented upon the symbolic significance of Wegg as a character who embodies dispersal and fragmentation, two major themes of *Our Mutual Friend* as a whole. Albert D. Hutter, for instance, identifies that a "problem" for several of the characters in Dickens's novel is "disarticulation": "characters are cut off from their work and from each other or like Wegg (at another extreme) from parts of themselves."[88] Elsewhere, Lawrence Frank suggests that "Silas Wegg's comic embodiment of the danger of dispersal, of fragmentation, is matched by his equally comic inclination to paralysis, to petrifaction … Inevitably, in the art of analogy Dickens so skilfully employs, Wegg's comic predicament comments upon the serious plights of other characters."[89] More recently, Alex Woloch argues that "Wegg's emblematic wooden leg doesn't only stand for both petrifaction and fragmentation but also stands *as* a product of the clash between embodiment and disembodiment that is produced *by* a character's standing for such abstractions in the first place."[90] Goldie Morgentaler, meanwhile, uses Wegg and his woodenness as an example through which to explore the theme of "mistaken and disembodied identity." *Our Mutual Friend*, Morgentaler writes, "is vitally concerned with issues of identity and with the difficulties of separating the genuine essence of an individual from its outer manifestation."[91] Unlike these critics, however, who are primarily concerned with the symbolic value of Wegg's wooden leg, Gavin "examines Wegg against the historical background of Dickens's interest in wooden legs, Victorian surgery and prosthetics, and nineteenth-century commodification of body parts."[92] I have continued Gavin's line of inquiry with specific regard to social class and the literary function of the prosthesis user in my blog post for the *Dickens* Our Mutual Friend *Reading Project*, where I argue that Dickens's representations of Wegg and the aptronymic Greenwich pensioner Gruff and Glum (who appears at the wedding of John Rokesmith and Bella Wilfer) draw our attention to "the link between physical and social mobility in the nineteenth century."[93] Below, I address the issue of class, a topic linked to both the materiality of the Wegg's prosthesis and his status as a physically reassembled Victorian male.

Readers first encounter Wegg as a character who is very much defined by his prosthesis. Before we become acquainted with Wegg's name,

Dickens describes him as "a man with a wooden leg," suggesting that his most distinguishing feature is the prosthesis that he uses.[94] At this point, Wegg operates at the lower end of the Victorian class system: he owns a stall that sells various miscellaneous items, including sweets, fruits, ballads, and gingerbread. Furthermore, Wegg claims to run errands for a nearby house, although, as the narrator notes, "he received such commissions not half-a-dozen times in a year, and then only as some servant's deputy."[95] Wegg's inability to attain a position as a servant emphasizes the privileging of physical wholeness that abounded in nineteenth-century society. Since Wegg has one leg, it is assumed that he is unable to perform household duties as well as a nondisabled employee. Wegg's false leg therefore stands for his inability to secure employment. He is rendered useless by social attitudes directed towards his amputation and use of a wooden leg. The leg stands for low income while metonymically it fits a stereotyped view of paupers who had for many years been represented as impaired or disfigured in one way or another. David Copperfield's sweetheart, Dora, for instance, associates such beggary with "a yellow face and a nightcap, or a pair of crutches, or a wooden leg, or a dog with a decanter-stand in his mouth, or something of that kind."[96] Artwork from the early modern period onwards supports this association of primitive prosthesis use with mendicancy and street work, as we can see in Figs. 4.1, 4.2, and 4.3.

Despite the symbolic status of his prosthesis, Wegg's wooden leg provides its user with certain advantages. In his line of work, which involves (for the most part) sitting and waiting for customers to arrive, Wegg's artificial leg is less of a hindrance than his healthy one. To keep his organic leg warm, he places it in a basket—an unusual choice of leg-warmer that in form closely resembles the "bucket" of a peg leg into which the user's stump would be inserted. Later, when he is asked if he likes his wooden leg, Wegg humorously responds, "Well! I haven't got to keep it warm," as though he's ready to be rid of the "good" leg too.[97] In this instance, Dickens light-heartedly echoes the words of journalist and fiction writer Charles Manby Smith, who, in his experimental literary essay "An Essay on Wooden Legs," mockingly proposed that limb-prosthesis wearers might possess advantages over bipeds: "look at the double risks of the double-footed, even in calamities that come unsought. The gout, that horrible visitant, has but half a victim in a one-legged man; of corns too he has but half a crop; his bunions never mar his quiet pilgrimage; and, come what may, he cannot by any possibility suffer from damp feet."[98] Considering the way that Wegg is described throughout the novel (as we

Fig. 4.1 Seventeenth-century etching of a beggar with two crutches and a wooden leg. Jacques Callot, *Beggar with a Wooden Leg*, c. 1622, etching with engraving, 13.8 × 0.8 cm, Wellcome Collection, London. Image courtesy of the Wellcome Collection. https://wellcomecollection.org/works/bssun4bb. CC BY 4.0

will see, he is as wooden in character as his prosthetic leg is in form), it is perhaps unsurprising that he chooses another wooden object to keep his real leg warm. Wegg's use of the basket suggests that he is comfortable using wooden objects as supplements to his physical form. The basket is not much different from the "bucket" part of his prosthesis that his stump goes into. Wegg's aptitude for prosthesis use is also implied by the way he uses his wooden leg in non-typical ways. Elsewhere, Wegg prods Venus with his wooden leg to discreetly gain his attention. In another instance, he is said to "take his wooden leg naturally."[99]

Ironically, Wegg's wooden leg helps him to secure a position as a literary man for Mr. Boffin, the recent inheritor of a considerable fortune. Boffin goes so far as to boast that his literary fellow is a prosthesis user. When John Rokesmith first approaches Boffin looking for employment, the Golden Dustman warns the young man that he has already made a recent appointment: "I have in my employment a literary man—*with* a wooden leg—as I have no thoughts of parting from" (emphasis

Fig. 4.2 Eighteenth-century etching illustrating an encounter between a peg-leg-using retired soldier and a pluralist. John Collier, *The Pluralist and Old Soldier*, 1770, etching with engraving, 22.4 × 23.2 cm, Wellcome Collection, London. Image courtesy of the Wellcome Collection. https://wellcomecollection.org/works/ztss4ux8. CC BY 4.0

Fig. 4.3 Early nineteenth-century etching of a hook-hand-using street hawker selling haddock. J.T. Smith, *Live Haddock*, 1815, etching, 18.8 × 11.3 cm, Wellcome Collection, London. Image courtesy of the Wellcome Collection. https://wellcomecollection.org/works/jeq6wsj2. CC BY 4.0

original).[100] In the light of Victorian notions of health and the corollary discourses of work and masculinity, it is odd to boast of employing someone with a physical impairment. Yet here the emphasis is on the prosthetic device rather than the missing limb. In this sense, Boffin's respect for

Wegg appears to be expounded by the amputee's use of a prosthetic limb. The fact that Wegg sells ballads and is physically impaired implies to Boffin that the amputee is well-read. Boffin employs Wegg to fulfil new social expectations that follow his newfound wealth, and Boffin feels it necessary to be learned and cultured in order to successfully perform his heightened social role. Since it was considered more respectable for amputees to use prostheses, Boffin judges Wegg in a favourable manner. To Boffin, Wegg's use of a wooden leg therefore signifies that he is genteel. A culturally aware Victorian reader would pick up on Boffin's assumptions, which are not only unusual but misinformed. As we have seen, the peg leg that Wegg wears associates him with maimed industrial workers, beggars, and naval veterans rather than members of the literati—though this latter group was not entirely without real-life associates with lower-limb difference (e.g. Lord Byron and, later, William Ernest Henley).

Despite Boffin's best intentions, comic attention is centred on his foolishness early in the novel. Not only is Boffin uninformed and completely taken in by Wegg's clumsy use of verse and false literary prowess (epitomized by Wegg's "wooden conceit and craft," which "kept exact pace with the delighted expectation of his victim"), but he also shows an ignorance towards popular contemporary connotations of peg legs, which were more commonly associated with beggary than respectability.[101] As Mihm suggests, in a time where the "tendency to equate external, bodily appearance with internal character" became popular and legitimized by scientific and medical doctrines, prosthesis type became an index for social value: "Prostheses from [the nineteenth century], far from being mere markers of technological progress, remain emblems, largely forgotten, of the demands posed by an 'age of appearances' in whose shadows we continue to live today."[102] An 1885 American etiquette manual suggested that "A man's walk" is "an index of his character and of the grade of his culture."[103] "The sight of a man, however respectably dressed," Mihm suggests, "hobbling down the street on an 'odious peg' would inevitably lead strangers to judge him in a negative light, as a 'cripple.'"[104] Thus, Dickens establishes Boffin as a man ignorant of this social code, since the Golden Dustman values Wegg more for having a wooden leg. Boffin's lack of awareness towards the social connotations of an unnatural-looking prosthetic device could be symptomatic of his own rapid social rise. In working-class social spaces, such as the factory, farm, or gin palace, there was much less impetus on the appearance and lifelike movement of a prosthetic limb.[105] Instead, the merit of an artificial body part was judged by utility:

if a wooden leg allowed an amputee to return to work, it was deemed successful. Justified by his quick rise from the lower end of the class spectrum, Dickens depicts Boffin as unaware of polite society's demands for more lifelike prostheses. For Wegg, however, Boffin's naivety is beneficial: it enables him to ascend through the social hierarchy.

Wegg also benefits from his apparent woodenness. By providing a contrast to the hegemonic norm of organic wholeness, the materiality of Wegg's prosthesis is reflected in the mannerisms, movements, and character of the wooden-leg user.[106] As the narrator observes:

> Wegg was a knotty man, and a close-grained, with a face carved out of a very hard material, that had just as much play of expression as a watchman's rattle. When he laughed, certain jerks occurred in it, and the rattle sprung. Sooth to say, he was so wooden a man that he seemed to have taken his wooden leg naturally, and rather suggested to the fanciful observer, that he might be expected—if his development received no untimely check—to be completely set up with a pair of wooden legs in about six months.[107]

Later, Wegg has a "hard-grained face," "stiff knotty figure," and is compared in looks to "a German wooden toy."[108] In another instance, he has "wooden countenance," and elsewhere a "wooden head."[109] Correspondingly, Wegg's name is an elision of "wooden" and "legg"—a splicing of language that signifies the merging of man and thing. As Carey suggests, Wegg is one of the most lucid representatives of Dickens's interest in "the border country between people and things":

> In a sense the wooden-legged men are at an intermediate stage of turning into wood, and with Silas Wegg the process has gone further. He is described as "knotty" and "close-grained," altogether so wooden that he seems to have grown his wooden leg naturally, and may be expected to develop a second one, Dickens conjectures, in about six months.[110]

In a manner not always consistent in Dickens's fiction, the characterization of Wegg is determined by his physical features.[111] In this case, Wegg's prosthesis is (in material if nothing else) the most visible signifier of the character's behaviour and inner qualities. To an almost comical degree, the narrator attributes the amputee with attributes akin to timber. That Wegg is "knotty" sets out that he is fixed and awkward, much like his wooden leg; that he is "close-grained" suggests that he is hard, obstinate, and narrow-minded. Stern-faced, with a mechanical laugh, Silas Wegg is in nearly all capacities presented as wooden by the narrator. Even the character's movements are described in terms of woodenness, since he prone to "stump" rather than walk. Unlike "the Oaken Lady" of Nathaniel Hawthorne's short story "Drowne's Wooden Image" (1844) (a wooden figurehead for a boat that is so mesmerizingly mimetic an effigy that she is thought to have come to life when her original comes to town), the aesthetic effect of Wegg's wooden artificial part is not only blatant, but all-consuming.[112] In appearance he is more wooden than the oak figure skilfully carved by the craftsman Drowne in Hawthorne's tale. As we learn, Wegg's prostheticized woodenness serves him well, for a time providing an imaginative alternative to organic physical wholeness.

The splicing of man and thing in Wegg also recalls the static, mechanized "hands" of the Victorian factory system. The human-machine dynamic of the factory was analysed in depth by Karl Marx in his contemporaneous critique of political economy, *Capital* (1867). As Tamara Ketabgian explains in her book *The Lives of Machines*, Marx's writing articulated how humans and machines jockeyed for dominance and subject status in the factory system: "the humans and machines of *Capital* shift restlessly between the role of host and prosthesis."[113] According to Marx, the factory arranged its workers as "parts of a living mechanism." The division of labour, Marx suggested, "mutilates the worker," transforming him into the "life-long organ of [a] partial function." Reversing the prosthetic function that technology is usually held to serve, Marx intimated that the worker, fragmented by monotonous routine, became a "living appendage of the machine."[114] Dickens's portrayal of a character in Silas Wegg, who one could argue is mechanized by his use of a prosthetic device, has parallels with the "human organs" of Marx's factory. In both Marx's and Dickens's respective writings, the dominance of the mechanical is brought to the fore. Whereas for Marx the human subject takes on the supposed passivity of the object/thing, Dickens gives thought to a more beneficial human-object relationship.

In *Our Mutual Friend*, Wegg's woodenness is simultaneously a help and hindrance to the amputee's social rise. Early on, when running a street stall, Wegg's hardness (which is transposed to the fruit that he sells) is somewhat off-putting for potential customers:

> Assuredly, this stall of Silas Wegg's was the hardest little stall of all the sterile little stalls in London. It gave you the face-ache to look at his oranges, the tooth-ache to look at his nuts. Of the latter commodity he had always a grim

> little heap, on which lay a little wooden measure which had no discernible inside, and was considered to represent the penn'orth appointed by Magna Charta.[115]

Here, stiffness is reflected not only in Wegg's character, but also in the produce that he attempts to sell. Moreover, his measures are fixed and rigid. Elsewhere, the very language with which the narrator describes Wegg's shrewdness draws from a lexical set connected with wood: "'Boffin will get all the eagerer for waiting a bit,' says Silas, *screwing* up, as he *stumps* along, first his right eye, and then his left. Which is something superfluous in him, for Nature has already *screwed* both pretty tight" (emphasis added).[116] Here both Wegg's facial expression and figurative composition—being "screwed" tight together—reinforce the view that the material properties of his prosthesis are reflected in his character. This transposition of qualities hardly seems negative in this instance, though, as Wegg is clearly someone who has his wits about him. The noisy and inarticulate way in which he is described to move, which again draws on his bodily state and use of prosthesis, similarly reflects his slow, methodical, and calculating nature.

These characteristics allow Wegg to infiltrate the Boffin family home and scheme the plan that almost collapses the world of the Golden Dustman. Wegg's wooden mannerisms are represented as directly responsible for his upward mobility when, after "stumping leisurely to the Roman Empire," Boffin offers his literary man with wooden leg a permanent position at the Bower, meaning that he can give up his street stall for a better living.[117] Revealing the advantages of a human-machine splice, Dickens brings into question the preference for organic physical completeness by presenting a temporarily successful, prostheticized, and conspicuously wooden alternative. Furthermore, Dickens collapses the social mandate for prostheses that allow users to pass. It is after all the recognition that he *is* a wooden-leg user that gains him employment in the first place.

However, Dickens's novel later reveals some unpleasant corollaries of Wegg's prosthesis use in the form of social marginalization and physical impairment. An incident that reveals the vulnerability of the amputee to unwanted attention occurs when Mr. Boffin callously asks Wegg how he acquired his wooden leg. Here, it is interesting that the Golden Dustman asks not how Wegg sustained his injury but rather how he obtained his prosthesis. Boffin's question can be read as both an attempt at subtlety and revelatory of his preoccupation with Wegg's false leg, which for him is both an object of fascination and a physical manifestation of literary knowledge and worldliness. The fact that Boffin makes this rather personal enquiry during his first meeting with Wegg once again emphasizes the former's ignorance towards the social protocols of bourgeois life. The inappropriateness of questions like this was mocked a few years earlier in *Household Words*. In a serialized chapter of George A. Sala's "The Great Hotel Question" (1856), the anonymous author ridicules Americans for their supposedly shameless inquisitiveness:

> There is but one instance on record, I believe, of a Yankee being worsted, in the query line of conversation; and this was the questioning Yankee who persisted in asking the dyspeptic man with a wooden leg how he had lost his missing leg, and after much pressing was told, on a solemn promise that he would ask no more questions, and under a penalty of dollars uncountable, that it had been bit off; whereupon, in an agony of uncertainty as to who or what had bitten it off, and how—whether it had fallen a victim to the jaws of deadly alligator, or catawampous panther, or fiercely-riled rattlesnake; and, fearing to break his word, or lose his dollars, he was crestfallen and confounded, and, ignominiously sloping, was seen no more in that territory.[118]

This passage reveals that even to the Victorians, who often displayed unsympathetic or suspicious attitudes towards disabled adults, enquiries like these were deemed inappropriate. What is particularly fascinating about this excerpt is that the American assumes that "bitten off" means that an animal inflicted the amputee's injury, entirely ruling out the very real possibility that the man may have lost his leg to the metallic jaws of the cotton-mill, paper-press, or another piece of dangerous machinery.[119]

Indelicate and invasive, Boffin's enquiry is symptomatic of his fast rise up the social scale. He asks a question that reflects ignorance of the feelings of others, which a Victorian reader might associate with an unrefined, working-class temperament, one standing at odds with his new social position. However, his enquiry also indicates the inconsiderate way that disabled people are often burdened by forthright, personal, and highly emotive questions in modern society.[120] Wegg's response to this enquiry implies that he is understandably offended: "Mr Wegg replied (tartly to this personal inquiry), 'In an accident.'"[121] This sharp retort suggests distress: Wegg neither enjoys being subjected to these enquiries nor is willing to go into detail about how he sustained his injury. As O'Connor has noted, lower-limb amputation was feared to have an emasculating effect on male patients: "It unmanned amputees, producing neurological disorders that gave the fragmented male body—or parts of it, anyway—a distinctly feminine side. Thrashing, twitching, and suffering from phantom pains, stumps showed a deep-rooted propensity for theatrical malingering that rivalled that of the hysterical herself."[122] In this case, the reproof that Wegg seems to suffer in response to Boffin's question intimates that the amputee feels that his masculinity is placed under scrutiny. The "tart" retort is a warranted defensive response and is discursively reinforced by the answer that he lost his leg in an accident. This reaction is intriguing since Boffin asked how Wegg got his wooden leg, not how he sustained his injury. It is therefore possible that Wegg asserts the cause to be an "accident" as a hasty defence of his masculinity: like the amputee who claims that his leg was bitten off, accident transforms the limb loss from a feminizing defect or troubling congenital deformity into a war wound or badge of honour—an emblem of masculine endeavour and national pride in the most extreme degree.[123]

Complicating Wegg's personal-injury narrative, Wegg's leg, as we later discover, was surgically amputated. It is also implied that the operation may have been carried out because of a congenital impairment since Mr. Venus, an experienced articulator of bones, reveals to Wegg that his amputated leg is abnormal: "You have got a twist in that bone, to the best of my belief. *I* never saw the likes of you." Venus goes on to remark that Wegg's amputated leg could be made use of as a "Monstrosity," further emphasizing its unusual form.[124] Assuming, then, that Wegg's leg was surgically removed because of complications arising from severe deformity rather than injury, his retort to Boffin's question provides evidence for the notion that it was less socially acceptable to be congenitally deformed than it was to lose a body part because of accident. *Our Mutual Friend* thus further complicates our understanding of physical wholeness and loss, since it suggests that the social position of the physically disabled could be contingent on how the subject lost her/his body part.

Dickens draws our attention to the ludicrous ends that a privileging of physical integrity can bring about when Wegg decides that to advance himself socially he must reunite with his lost leg. Going to Mr. Venus's shop with the intention of purchasing the remains of his amputated leg, Wegg explains: "I shouldn't like—I tell you openly I should not like—under such circumstances, to be what I may call dispersed, a part of me here, and a part of me there, but should wish to collect myself like a genteel person."[125] Wegg takes the contemporary privileging of physical wholeness to its logical extreme. The perceived link between social mobility and physical completeness here is key: Wegg believes that one must be fully intact to make progress in the world. To match his new, elevated social position, he must be whole. But in an absurd and cruelly humorous manner Dickens shows that a wooden leg is not enough to make a man feel whole again after losing a leg. The narrator also draws the reader's attention to the conflicting cultural messages regarding physical integrity: one should strive to maintain an appearance of completeness but to artificially cultivate one is fraudulent. Through Wegg, Dickens show us that conflicting demands can bring a man to a state of confusion.

Elsewhere, Wegg accentuates the flaws in his prosthesis. In these instances, Dickens can be read to actively engage with contemporary debates surrounding what constituted "a limb which shall be presentable in polite society."[126] Most luridly, Wegg describes to Venus how he would like to see wooden legs adapted: "Mr Wegg next modestly remarks on the want of adaption in a wooden leg to ladders and such-like airy perches, and also hints at the inherent tendency in that timber fiction, when called into action for the purposes of a promenade on an ashy slope, to stick itself into yielding foothold, and peg its owner to one spot."[127] Here, Wegg reveals the practical limitations of his prosthesis in terms of manual work outdoors. By drawing attention to the wooden leg's unsuitability for climbing ladders and its tendency to get stuck in soft ground, Wegg outlines the inhuman and unwieldy nature of wooden prostheses, seemingly reinforcing the contemporary preference for organic wholeness, which as we have seen was not a consistent position throughout the novel. Wegg's difficulty ascending ladders suggests that the wooden prosthesis is unable to mimic the full range of movement and stability provided by a natural leg. It also implies that, unlike an organic leg, a wooden one is difficult to manoeuvre, meaning that it is in some ways independent of, rather than integrated within, the body. The wooden prosthesis's propensity for getting stuck underlines how these devices in no way successfully mimicked the size and shape of a human foot. Since timber is often heavy and inflexible, prosthesis makers tended not shape their prosthesis to the form and dimensions of a real leg. Instead, as we have seen, peg legs were thin and rolling-pin shaped. In *Our Mutual Friend*, Wegg promulgates the flaws in limb prosthesis that primarily stem from their inorganic nature. While Wegg is represented as a man who is as wooden as his prosthesis, his device is depicted as flawed, just as he is as a character. On the one hand, then, the depiction of Wegg seems to support the hegemony of organic wholeness as its alternative, prosthetic woodenness, is debunked. But, on the other hand, this process brings into question the impulse to prostheticize, a desire encouraged by a cultural privileging of physical integrity.

The faults that Wegg identifies in his prosthesis become most problematic to the amputee while he sneakily scours Mr. Boffin's dust mounds to find evidence that could lead to his master's downfall. In several ways, Wegg's prosthesis and its material influence can be read as the primary causes of his own fall from grace. It is, after all, because Wegg is unable to find Mr. Harmon's most recent will before Boffin that the amputee ultimately fails. Likewise, it is Wegg's wooden obstinacy and general "knotty" demeanour that stimulate his greed in the first place. Boffin identifies earlier in the novel that "a literary man—*with* a wooden leg—is liable to jealousy," yet little is he aware that it is him and his fortune that Wegg is most jealous of.[128]

In several senses, Wegg can be understood in connection with the "disabled male dichotomy" that Holmes identifies in *Fictions of Affliction*. Holmes suggests that there is a "representational gap" between the portrayal of disabled men and that of boys in Victorian literature in which the latter evoke "emotional excess as the intensity of pure pathos," while the former represent "the excess of bilked emotion, imposture, and inauthenticity."[129] Wegg is depicted as a fraud—to a rather comical extent. Boffin's assumption that the amputee is a literary man is shown as well wide of the mark. Though the Golden Dustman knows no better (Wegg's "wooden conceit and craft" are said to keep "exact pace with the delighted expectation of his victim"), the amputee's knowledge is proven to be farcical very early on. A clear indicator of Wegg's façade becomes apparent when Boffin asks his literary man what the difference is between the "Rooshan" and Roman Empire. Here, Wegg retorts: "The difference, sir? … The difference, sir? There you place me in a difficulty, Mr Boffin. Suffice it to observe, that the difference is best postponed to some other occasion when Mrs Boffin does not honour us with her company. In Mrs Boffin's presence, sir, we had better drop it."[130] Earlier, after having been offered employment by Boffin, Wegg is described in the following way:

> His gravity was unusual, portentous, and immeasurable, not because he admitted any doubt of himself, but because he perceived it necessary to forestall any doubt of himself in others. And herein he ranged with that very

> numerous class of impostors, who are quite as determined to keep up appearances to themselves, as to their neighbours.[131]

Wegg is thus singled out as a fraud and associated with the swindling beggars described in Mayhew's *London Labour and London Poor*, even before the amputee attempts to extort Boffin of his riches. Like beggars that use an impression of disability for profit-making purposes, the depiction of Wegg troublingly elides the physically incomplete body with duplicitous schemes.

Related to the theme of fraudulence, Dickens brings the falsity of Wegg's wooden leg to the fore late in the novel, where we see that prostheses are a kind of sham. The discrepancy between flesh and artificial body part is depicted when Wegg is described as worn and haggard by his endeavours to grow rich at Boffin's expense. As the narrator notes: "So gaunt and haggard had he grown at last, that his wooden leg showed disproportionate, and presented a thriving appearance in contrast with the rest of the plagued body, which might almost have been termed chubby."[132] In this instance, the fact that Wegg's prosthesis remains unaltered by the stress occasioned by the wearer's pursuit of wealth provides a contrast with the body of the amputee, which is described as vulnerable to the symptoms of stress. Despite the earlier descriptions of Wegg's complete woodenness, the sharp divide between hard, unchanging substance and soft, vulnerable human tissue becomes apparent. Like the "hands" of the Victorian factory, whose physical discrepancy to the unrelenting force of industrial machinery routinely resulted in the kinds of accidents reported and lamented by commentators such as Henry Morley, Wegg's organic body pales in comparison to his artificial wooden leg.[133] Unfortunately for Wegg, the apparent strength and social mobility occasioned by his woodenness proves a façade: not only does his health suffer because of his designs, but the designs themselves also fail, since it turns out he failed in obtaining the final will written by Harmon senior. Though Wegg remains wooden-looking, obstinate, and greedy until the end of the novel, his woodenness no longer appears an advantage. Wegg's use of prosthesis is one of the very flaws that has led to his plan failing: the nondisabled Boffin navigated the dust heap better than the physically impeded amputee. Fluctuating between a position that either mocks the hegemony of organic physical wholeness or reaffirms it, Wegg's fall from grace sides with the latter.

In addition to Wegg, Dickens presents the reader with a second prosthesis-wearing amputee in *Our Mutual Friend*, one whose social mobility, like the wooden villain, is linked to his use of prosthesis: the navy veteran Gruff and Glum, who appears in Book 4, Chap. 4.[134] Unlike Wegg, Gruff and Glum is a double amputee. The representation of this character more straightforwardly equates physical mobility to social mobility. Here, the double-leg amputee is represented as a character whose lower-limb prostheses restrict him to the social space of retirement: Greenwich hospital—Britain's home from the seventeenth century for its disabled navy veterans. The very name Gruff and Glum indicates the character of the man (a hardened navy veteran) and hints towards the psychological impact of limb amputation. Prior to Bella's arrival in town (for her wedding with John Rokesmith), Gruff and Glum is said to have "no other object in life but tobacco" and is described as "[s]tranded … in a harbour of everlasting mud."[135] The description is both literal and metonymic: harbours, such as Greenwich's were often muddy, meaning that his physical mobility was restricted by the limitations placed upon him by his prostheses, which were not well-suited for walking on soft ground. Still, he is also trapped by his retired status and the symbolic meaning attached to his wooden legs. Oddly, the veteran's body is reawakened by his contact with Bella: "For years, the wings of his mind had gone to look after the legs of his body; but Bella had brought them back for him per steamer, and they were spread again."[136] Here, the presence of Bella liberates Gruff and Glum's thoughts from his injuries, underscoring the importance of human contact within the rehabilitation process of an amputee. In this regard, Gruff and Glum's portrayal is a comment on the social segregation experienced by patients of institutions such as Greenwich Hospital.[137] Like Wegg, the woodenness of Gruff and Glum's prostheses is also reflected in his character. His artificial legs stand as visual signifiers for his former profession that remain a strong influence upon him. In response to a compliment from Bella, using a naval idiom, "Gruff and Glum … wished her ji and the fairest of fair wind and weather; further, in a general way requesting to know what cheer? And scrambling up on his two wooden legs to salute, hat in hand, ship-shape, with the gallantry of a man-of-warsman and a heart of oak."[138] As the narrator's description suggests, he has become a half-wooden masthead, like the wooden midshipman in *Dombey and Son*. Here, Gruff and Glum's choice of language directly draws from his former days at sea, while the description that he has "a heart of oak" suggests that he is akin both in character and in substance to a naval vessel.

Like a retired ship, he is worn, wooden and moored to a dock, where it is destined to remain until his eventual demise. Like Wegg, Gruff and Glum is pegged to a low social standing by prosthesis.[139]

Wegg's Legacy

In 1877, seven years after Dickens's death, an article appeared in *All the Year Round* (at this time commissioned by Charles Dickens Jr.) that reaffirmed some of the associations with inauthenticity, dishonesty, and imposture that Wegg embodies in *Our Mutual Friend*. In "Mr. Wegg and his Class," the author unsympathetically discussed the imprudence of street beggars who were "engaged in the crossing-sweeping line of business," and were the real-life equivalents of Dickens's fictional amputee.[140] Of the supposedly deplorable figures that the author describes, one in particular is said to be uncannily similar to Wegg: "Like his great prototype, he had a wooden leg; like him he was literary; and, finally, like him, under cover of affecting to follow his profession, he assiduously cultivated another, namely that of Humbug."[141] Here, then, not only is this real-life figure also an amputee peg-leg wearer, but he is also considered a literary man and depicted as unscrupulous. Like Wegg and the disabled street workers described by Mayhew, the beggar deployed a number of duplicitous strategies for financial gain. He used his wooden leg to inspire the idea that he was a war veteran. He drew upon the sympathy of others, claiming to be in constant pain. After his wooden leg broke, he used this as an excuse to demand extra money from passers-by. And he claimed to have found Salvation and read Christian verses aloud to encourage charitable donations.

What the article fails to consider, and what *Our Mutual Friend* investigates with much more nuance and complexity, however, are the limited options available to lower-class amputees. It was no doubt easy enough to be annoyed at wooden-leg users, such as Wegg's real-life progeny, for their duplicitous strategies. But with inflexible and widely held views that restricted working opportunities for impaired individuals, trickery provided rare (albeit limited) opportunities for financial success. It was not as if the apparently wondrous enabling devices of the contemporary prosthesis market were available to street amputees. The devices that they could afford, wooden pegs, often failed to provide their users with the ability to perform even basic physical feats. Sometimes these devices even made the physical situation of their users worse, leaving street work or begging as some of the only viable options. It was thus a vicious cycle for lower-class amputees. While those who could afford "artificial legs" might facilitate social mobility once more—including devices promised to enable them to pass as normal—those who could not were often rendered both physically and socially immobile. These inevitabilities, brought about by a culture that privileged wholeness at the expense of those missing body parts, were challenged by narratives including *Our Mutual Friend*, which experimented with the idea that amputees could be social climbers. It is telling, though, that Dickens's successors saw Wegg as the archetypal "peg-legged beggar" rather than a transgressive social climber. What *Our Mutual Friend* and "Mr Wegg and His Class" share is a vision that those who are oppressed by a society unwilling to employ or respect them might cultivate a ruthless streak, which could manifest in behaviour akin to Wegg's. This attitude has clear ableist and disablist underpinnings. It deflects blame away from society by criminalizing the marginalized subject. But both sources, like the other fictional works discussed in this chapter, bring into question the premium placed on physical completeness, since they draw attention to the counterintuitive results that normative bodily preferences might give rise to.

This chapter has shown how writers such as Dickens experimented with the social trajectories of their fictional prosthesis users in ways that challenged but did not ultimately refute the contemporary privileging of physical wholeness. By analysing these writings through a disability-studies lens, self-contradictions in the philosophy of prosthetically supplementing the human body come to the fore. If the body is sacrosanct, its imitation, a process forced upon those whose bodies are perceived to be incomplete, becomes stigmatized as counterfeit. Thus, the very desire to replicate the body brings about its own critique, since, if a prosthetic part made is a poor replica (as in the case of peg legs), its purpose is defunct. Then again, if it made perfect (as supposedly was the case with high-end artificial limbs), it becomes an emblem of deceit. But also, critical to the social element of this paradox, the more superior devices made to hide physical losses come at too high a price for those who arguably need them the most—those from the lower end of the class system, whose jobs require physical mobility. The inevitable result is that social mobility becomes impossible for amputees at the lower end of the class spectrum. Ironically, owing to the very hegemony of physical wholeness, feigning a physical difference becomes profitable for those at the bottom as this condition evokes pity, and thereby increases potential for charitable gain. By transgressively imagining prosthesis users who because of their integration with a primitive prosthesis can elevate their social position, the representations discussed upend cultural preferences. Even though works such as *Our Mutual Friend* end conservatively, reifying the premium placed on wholeness, this act of self-assurance reveals the fragility of a concept that invited interrogation and required so much cultural buttressing.

Notes

1. Ibid., 159.
2. Ibid.
3. For more on the idea of Victorian technology as a kind of prosthesis, see Tamara Ketabgian, *The Lives of Machines: The Industrial Imaginary in Victorian Literature and Culture* (Ann Arbor: University of Michigan Press, 2011), 19–29.
4. Charles Dickens, *Hard Times* (London: Penguin, 2003).
5. Stephen Mihm, "'A Limb Which Shall Be Presentable in Polite Society': Prosthetic Technologies in the Nineteenth Century," in *Artificial Parts, Practical Lives: Modern Histories of Prosthetics*, eds. Katherine Ott, David Serlin, and Stephen Mihm (New York: New York University Press, 2002).
6. Erin O'Connor, "'Fractions of Men': Engendering Amputation in Victorian Culture," *Comparative Studies in Society and History* 39, no. 4 (1997); Erin O'Connor, *Raw Material: Producing Pathology in Victorian Culture* (Durham: Duke University Press, 2000).
7. [William Blanchard Jerrold], "Eyes Made to Order," *Household Words* 4, no. 81 (1851): 64.
8. Charles Dickens, *Our Mutual Friend* (London: Everyman, 2000).
9. O'Connor, *Raw Material*, 104.
10. Cindy LaCom, "'The Time Is Sick and Out of Joint': Physical Disability in Victorian England," *PMLA* 120, no. 2 (2005): 547. Also see Michael Oliver, *The Politics of Disablement* (Basingstoke: Macmillan Education, 1990), 25–42.
11. LaCom, "The Time," 548.
12. Qtd. in Robin M. Reeve, *The Industrial Revolution, 1750–1850* (London: University of London Press, 1971), 186.
13. Marta Russell, *Beyond Ramps: Disability at the End of the Social Contract* (Monroe: Common Courage Press, 1998), 59.
14. Jamie L. Bronstein, *Caught in the Machinery: Workplace Accidents and Injured Workers in Nineteenth-Century Britain* (Stanford University Press, 2008), 96.
15. Ibid., 88.

16. Qtd. in ibid.
17. Mihm, "A Limb," 288.
18. Auguste Boissonneau, *General Observations on Artificial Eyes, Their Adaption, Employment and the Means of Procuring Them* (Paris: J.-B. Baillière and Son, 1862), 7.
19. Martha Stoddard Holmes, *Fictions of Affliction: Physical Disability in Victorian Culture* (Ann Arbor: University of Michigan Press, 2004).
20. Ibid.
21. David Owen, *English Philanthropy, 1660–1960* (Cambridge: Belknap Press of Harvard University Press, 1964), 217.
22. Holmes, *Fictions*, 119.
23. Susan M. Schweik, *The Ugly Laws: Disability in Public* (New York: NYU Press, 2009), 111.
24. Henry Mayhew, *London Labour and the London Poor: A Cyclopaedia of the Condition and Earnings of Those That Will Work, Those That Cannot Work, and Those That Will Not Work*, Vol. 4 (London: Griffin, Bohn, and Company, 1862).
25. Ibid., 429.
26. Victor Hugo, *Notre-Dame de Paris*, trans. Alban Krailsheimer (Oxford: Oxford University Press, 2009).
27. Arthur Conan Doyle, "The Man with the Twisted Lip," *Strand Magazine: An Illustrated Monthly* 2 (1891).
28. Irina Metzler writes that as early as 806 a Carolingian capitulary issued at Nimwegen forbid almsgiving to beggars capable of working with their hands and around 820 Louis the Pious ordered supervisors to monitor beggars so that simulators could not hide among them. Irina Metzler, *A Social History of Disability in the Middle Ages: Cultural Considerations of Physical Impairment* (London: Routledge, 2013), 169.
29. Henry Mayhew, *London Labour and the London Poor: A Cyclopaedia of the Condition and Earnings of Those That Will Work, Those That Cannot Work, and Those That Will Not Work*, vol. 1 (London: Griffin, Bohn, and Company, 1861), 330–59.
30. Ibid., 330.
31. Ibid.
32. John Brown, "A Memoir of Robert Blincoe, an Orphan Boy," in *Factory Lives: Four Nineteenth-Century Working-Class Autobiographies*, ed. James R. Simmons Jr. (London: Broadview, 2007). William Dodd, *A Narrative of the Experience and Sufferings of William Dodd, A Factory Cripple. Written by Himself*, in *Factory Lives: Four Nineteenth-Century Working-Class Autobiographies*, ed. James R. Simmons, Jr. (London: Broadview, 2007). Blincoe's narrative does not fit within a traditional conception of autobiography as it was written by the journalist John Brown, though

Janice Carlyle makes the case for us to consider it as one due to expanding conceptions of the genre wrought by poststructuralist analysis and historical investigations of self-life-writing by women and ethnic minorities. Janice Carlyle, "Introduction," in *Factory Lives: Four Nineteenth-Century Working-Class Autobiographies*, ed. James R. Simmons Jr. (London: Broadview, 2007), 29–39.

33. Dodd, *A Narrative*, 221.
34. Mayhew, *London Labour*, vol. 1, 330.
35. See Holmes, *Fictions*, 94–132.
36. O'Connor, *Raw Material*, 130.
37. Boissonneau, *General Observations*, 7.
38. Auguste Boissonneau, *Method of Complete and Individual Appropriation of Artificial Eyes, Comprising the Different Kinds of Advertisements with the Wood Cuts Belonging to Them: M. Auguste Boissonneau's Moveable Artificial Eyes* (London: W. T. Soulby, [1853]), 3.
39. Frederick Gray, *Automatic Mechanism, as Applied in the Construction of Artificial Limbs, in Cases of Amputation* (London: H. Renshaw, 1855), 46 and 55.
40. Qtd. in ibid., 182.
41. Vanessa Warne, "'To Invest a Cripple with Peculiar Interest': Artificial Legs and Upper-Class Amputees at Mid-Century," *Victorian Review* 35, no. 2 (2009): 84.
42. Gray, *Automatic*, 94.
43. Oliver Wendell Holmes, "The Human Wheel, Its Spokes and Felloes," in *Sounds from the Atlantic*, by Oliver Wendell Holmes (Boston: Ticknor and Fields, 1864), 307.
44. Ibid., 306–307.
45. "Legs: Wooden and Otherwise." *All the Year Round* 14, no. 350 (1875): 463.
46. Ibid., 464.
47. "Street Sights in Constantinople," *All the Year Round*, no. 38 (1860): 279.
48. Exceptions to this trend include Thomas Hudson's "The Cork Leg" ([1819–1844?]) and Frances Parker's *The Flying Burgermaster: A Legend of the Black Forest* (1832). As I have shown in Chap. 2, the sophisticated devices depicted are imagined to have wills of their own, making them dangerous for their users. [Thomas Hudson], "The Cork Leg," Broadside Ballads Online, accessed June 21, 2018, http://ballads.bodleian.ox.ac.uk/view/edition/5413; [Frances Parker], the Countess of Morley, *The Flying Burgermaster: A Legend of the Black Forest* (n.p.: F. Morley, 1832).
49. Bronstein, *Caught*, 8–9.

50. E. J. Chaloner, H. S. Flora, and R. J. Ham, "Amputations at the London Hospital 1852–1857," *Journal of the Royal Society of Medicine* 94, no. 8 (2001): 409. Also see O'Connor, "Fractions," 744–46; Roger Cooter and Bill Luckin, *Accidents in History: Injuries, Fatalities and Social Relations* (Amsterdam: Rodopi, 1997); and John Williams-Searle, "Courting Risk: Disability, Masculinity, and Liability on Iowa's Railroads, 1868–1900," *The Annals of Iowa* 58, no. 1 (1999).
51. Dodd, *A Narrative*, 210.
52. Gray, *Automatic*, 107.
53. Gray notes that there were a few exceptions to this trend: some employers, including Great Western Railway had a special prosthesis fund for its employees. Gray, *Automatic*, 107. The North Western railway company had similar provisions in place. John Pendleton, "A Great Railway," *Good Words*, January 1898, 477.
54. George E. Marks, *Marks' Patent Artificial Limbs with Rubber Hands and Feet* (New York: A. A. Marks, 1888), 47.
55. While, upon close observation, even the most high-tech artificial legs were quite patently made of wood, they were designed to be worn under clothes, thus disguising their non-human form. Their ability to emulate the natural gait of a real leg meant that, while concealed, they often went unperceived. Because of their design, peg legs either protruded beyond the natural contours of the stump or were not thick enough to fill trouser legs, meaning that they could not be easily worn under garments. They were, therefore, often exposed for all to see.
56. Harvey Green, *Wood: Craft, Culture, History* (London: Penguin, 2007), xxv.
57. "Different Uses of Wood in the Arts," *The London Journal, and Weekly Record of Literature, Science, and Art* 36, no. 916 (1862): 136.
58. Green, *Wood*, xxv.
59. David Edgerton, *The Shock of the Old: Technology and Global History since 1900* (London: Profile, 2008).
60. Adrienne E. Gavin, "Dickens, Wegg, and Wooden Legs," *Our Mutual Friend*: The Scholarly Pages, accessed June 18, 2018, http://omf.ucsc.edu/london-1865/victorian-city/wooden-legs.html.
61. John Carey, *The Violent Effigy: A Study of Dickens' Imagination* (London: Faber and Faber, 1973), 91.
62. Charles Dickens, *Dombey and Son* (London: Penguin, 2002); Herbert Sussman and Gerhard Joseph, "Prefiguring the Posthuman: Dickens and Prosthesis," *Victorian Literature and Culture* 32, no. 2 (2004): 619.
63. Charles Dickens, *The Pickwick Papers* (Clinton: Colonial Press, 1868); Charles Dickens, *Nicholas Nickleby*, 2 vols. (London: Scholar Press, 1982); Charles Dickens, *The Old Curiosity Shop* (London: Chapman &

Hall, 1841); Charles Dickens, *Barnaby Rudge* (Oxford: Oxford University Press, 2003); Charles Dickens, *Martin Chuzzlewit*, 4 vols. (New York: Hurd and Houghton, 1869); Charles Dickens, *David Copperfield* (London: Penguin, 2004).

64. Dickens, *Nicholas Nickleby*, 157.
65. Dickens, *The Pickwick*, 438.
66. Dickens, *The Old*, 119.
67. Dickens, *Barnaby*, 656.
68. Dickens, *Martin Chuzzlewit*, vol. 2, 282.
69. Dickens, *David Copperfield*, 97.
70. Ibid., 271. Thomas Hood, *Miss Kilmansegg and Her Precious Leg*, in *Selected Poems of Thomas Hood*, ed. John Clubbe (Cambridge: Harvard University Press, 1970). Also see V. R. "The Wooden Legs in Dickens," *Notes and Queries* 171 (1936).
71. Charles Dickens, *The Mudfog Papers* (New York: J. W. Lovell Company, 1883), 628.
72. Jay Clayton, "Hacking the Nineteenth Century," in *Victorian Afterlife: Postmodern Culture Rewrites the Nineteenth Century*, eds. John Kucich and Dianne F. Sadoff (London: University of Minnesota Press, 2000), 189.
73. Charles Dickens, *The Selected Letters of Charles Dickens*, ed. Jenny Hartley (Oxford: Oxford University Press, 2012), 194.
74. [Richard H. Horne], "Dust; or Ugliness Redeemed," *Household Words* 1, no. 16 (1850): 379–84; [Charles Dickens], "New Year's Day," *Household Words* 19, no. 458 (1859): 97–102; [Sarah Smith], "The Lucky Leg," *Household Words* 19, no. 469 (1859): 374–80.
75. For scholarship that attempts to explain Dickens's interest in prostheses, see Gavin, "Dickens." Also see Michael Cotsell, *The Companion to Our Mutual Friend* (London: Allen and Unwin, 1986), 50–51; and Michael Allen, *Charles Dickens's Childhood* (Basingstoke: MacMillan, 1988), 16 and 77.
76. Horne, "Dust," 379.
77. Karl Marx, *Capital: A Critical Analysis of Capitalist Production*, trans. Samuel Moore and Edward Aveling (London: Kimble & Bradford, 1938).
78. George Henry Lewes, "Realism and the Art of the Novel," in *Literary Criticism of George Henry Lewes*, ed. Alice R. Kaminsky (Lincoln: University of Nebraska Press, 1964), 98.
79. As Caroline Evans notes, this trope of Dickens's fiction is also made manifest, yet also complicated, by the portrayal of Jenny Wren, her doll-making business and the confused way in which she refers to people as dolls and dolls as people. Caroline Evans, *Fashion at the Edge: Spectacle, Modernity and Deathliness* (London: Yale University Press, 2003), 167.

80. Dorothy Van Ghent, *The English Novel: Form and Function* (New York: Harper and Row, 1953), 158–59.
81. Samuel Smiles, *Self-Help; with Illustrations of Character and Conduct* (London: John Murray, 1859), 240.
82. Ibid., 246.
83. Ibid., 245.
84. John Ruskin, *Modern Painters*, vol. 5 (London: George Allen, 1906), 289.
85. James Eli Adams, *Dandies and Desert Saints: Styles of Victorian Masculinity* (Ithaca: Cornell University Press, 1995), 150.
86. Donald E. Hall, ed., *Muscular Christianity: Embodying the Victorian Age* (Cambridge: Cambridge University Press, 1994), 7.
87. Charles Kingsley, "The Science of Health," in *Sanitary and Social Lectures and Essays*, by Charles Kingsley (n.p.: ReadHowYouWant, 2008), 34.
88. Albert D. Hutter, "Dismemberment and Articulation in Our Mutual Friend," *Dickens Studies Annual* 11 (1983): 154.
89. Lawrence Frank, *Charles Dickens and the Romantic Self* (Lincoln: University of Nebraska Press, 1984), 27.
90. Alex Woloch, *The One* vs. *the Many: Minor Characters and the Space of the Protagonist in the Novel* (Princeton: Princeton University Press, 2003), 354n2.
91. Goldie Morgentaler, "Dickens and the Scattered Identity of Silas Wegg," *Dickens Quarterly* 22, no. 2 (2005): n.pag.
92. Gavin, "Dickens."
93. Ryan Sweet, "Legs of Wegg and Others," Dickens *Our Mutual Friend* Reading Project, last modified, August 5, 2015, https://dickensourmutualfriend.wordpress.com/2015/08/05/month-16-august-1865-legs-of-wegg-and-others/.
94. Dickens, *Our Mutual*, 46.
95. Ibid., 47.
96. Dickens, *David Copperfield*, 545–46.
97. Dickens, *Our Mutual*, 50.
98. Charles Manby Smith, "An Essay on Wooden Legs, with Some Account of Herr Von Holtzbein," *Tait's Edinburgh Magazine* 21, no. 244 (1854): 231.
99. Dickens, *Our Mutual*, 513–15. For more on devices that were imagined to enhance their users, see Chap. 3.
100. Dickens, *Our Mutual*, 103.
101. Ibid., 60.
102. Mihm, "A Limb," 294.
103. Qtd. in ibid., 288.
104. Ibid., 289.

105. Katherine Ott makes a similar point regarding artificial eyes. Katherine Ott, "Hard Wear and Soft Tissue: Craft and Commerce in Artificial Eyes," in *Artificial Parts, Practical Lives: Modern Histories of Prosthetics*, eds. Katherine Ott, David Serlin, and Stephen Mihm (New York: New York University Press, 2002), 152–53.
106. For more on this topic, also see Sweet, "Legs."
107. Dickens, *Our Mutual*, 48.
108. Ibid., 524.
109. Ibid., 624 and 698.
110. Carey, *The Violent*, 101 and 103.
111. See Paul Marchbanks, "From Caricature to Character: The Intellectually Disabled in Dickens's Novels," *Dickens Quarterly* 23 (2006).
112. Nathaniel Hawthorne, "Drowne's Wooden Image," in *Mosses from an Old Manse*, by Nathaniel Hawthorne, vol. 2 (Boston: Ticknor and Fields, 1854).
113. Ketabgian, *The Lives*, 19.
114. Marx, *Capital*, 548, 482, 458, and 614.
115. Dickens, *Our Mutual*, 48.
116. Ibid., 80.
117. Ibid., 196.
118. [George A. Sala], "The Great Hotel Question, in Three Chapters; Chapter the Third," *Household Words* 13, no. 310 (1856): 148.
119. Related to Dickens's interest in America and Americans, discussions of the American Civil War in *All the Year Round* provide important context to the author's anxieties about the dissolution of class barriers. See "Princely Travel in America," *All the Year Round* 8, no. 184 (1862); and Kylee-Anne Hingston, "'Skins to jump into': The Slipperiness of Identity and the Body in Wilkie Collins's *No Name*," *Victorian Literature and Culture* 40, no. 1 (2012).
120. See, for example, Alison Kafer, *Feminist, Queer, Crip* (Bloomington: Indiana University Press, 2013), 43; and Sunaura Taylor, *Beasts of Burden: Animal and Disability Liberation* (London: The New Press, 2017), 134–48.
121. Dickens, *Our Mutual*, 50.
122. O'Connor, *Raw Material*, 104.
123. For more on the bitter-sweet, conflicted views on missing limbs as badges of honour or courage in the context of the nineteenth century, see Lisa Herschbach, "Prosthetic Reconstruction: Making the Industry, Re-Making the Body, Modelling the Nation," *History Workshop Journal* 44 (1997); and Jalynn Olsen Padilla, "Army of "Cripples": Northern Civil War Amputees, Disability, and Manhood in Victorian America" (PhD diss., University of Delaware, 2007).

124. Dickens, *Our Mutual*, 82–87.
125. Ibid., 87.
126. Holmes, "The Human," 307.
127. Dickens, *Our Mutual*, 323.
128. Ibid., 192.
129. Holmes, *Fictions*, 98 and 95.
130. Dickens, *Our Mutual*, 60–62.
131. Ibid., 56.
132. Ibid., 830.
133. [Henry Morley], "Ground in the Mill," *Household Words* 9, no. 213 (1854).
134. Also see Sweet, "Legs."
135. Dickens, *Our Mutual*, 707.
136. Ibid.
137. This critique stands at odds to Dickens's usual championing of public hospitals. See Katharina Boehm, *Charles Dickens and the Sciences of Childhood* (Basingstoke: Palgrave, 2013), 79–111; and Louise Penner, "Dickens, Metropolitan Philosophy and the London Hospitals," *Victorian Medicine and Popular Culture*, eds. Louise Penner and Tabitha Sparks (London: Routledge, 2015).
138. Dickens, *Our Mutual*, 710.
139. Dickens's other prosthesis-using navy veteran, Captain Cuttle from *Dombey and Son*, is also shown to be relatively socially static, though his status is raised slightly at the end of the novel when he becomes a recognized business partner to Solomon Gill, owner of The Midshipman shop.
140. "Mr Wegg and His Class," *All the Year Round* 18, no. 441 (1877): 250–53. Smith earlier emphasized the prolificacy of maimed crossing-sweepers in *Curiosities of London Life*. Charles Manby Smith, *Curiosities of London Life; or, Phases, Physiological and Social, of the Great Metropolis* (London: W. and F. G. Cash, 1857), 54.
141. "Mr Wegg," 250–51.

References

Adams, James Eli. *Dandies and Desert Saints: Styles of Victorian Masculinity*. Ithaca: Cornell University Press, 1995.

Allen, Michael. *Charles Dickens's Childhood*. Basingstoke: Macmillan, 1988.

Boehm, Katharina. *Charles Dickens and the Sciences of Childhood*. Basingstoke: Palgrave, 2013.

Boissonneau, Auguste. *Method of Complete and Individual Appropriation of Artificial Eyes, Comprising the Different Kinds of Advertisements with the Wood*

Cuts Belonging to Them: M. Auguste Boissonneau's Moveable Artificial Eyes. London: W. T. Soulby, [1853].
———. *General Observations on Artificial Eyes, Their Adaption, Employment and the Means of Procuring Them.* Paris: J.-B. Baillière and Son, 1862.
Bronstein, Jamie L. *Caught in the Machinery: Workplace Accidents and Injured Workers in Nineteenth-Century Britain.* Stanford University Press, 2008.
Brown, John. "A Memoir of Robert Blincoe, an Orphan Boy." In *Factory Lives: Four Nineteenth-Century Working-Class Autobiographies*, edited by James R. Simmons Jr, 87–180. London: Broadview, 2007.
Carey, John. *The Violent Effigy: A Study of Dickens' Imagination.* London: Faber and Faber, 1973.
Carlyle, Janice. "Introduction." In *Factory Lives: Four Nineteenth-Century Working-Class Autobiographies*, edited by James R. Simmons Jr., 11–76. London: Broadview, 2007.
Chaloner, E. J., H. S. Flora, and R. J. Ham. "Amputations at the London Hospital 1852–1857." *Journal of the Royal Society of Medicine* 94, no. 8 (2001): 409–12.
Clayton, Jay. "Hacking the Nineteenth Century." In *Victorian Afterlife: Postmodern Culture Rewrites the Nineteenth Century*, edited by John Kucich and Dianne F. Sadoff, 186–210. London: University of Minnesota Press, 2000.
Cooter, Roger, and Bill Luckin. *Accidents in History: Injuries, Fatalities and Social Relations.* Amsterdam: Rodopi, 1997.
Cotsell, Michael. *The Companion to Our Mutual Friend.* London: Allen and Unwin, 1986.
Davis, Lennard J. *Bending Over Backwards: Disability, Dismodernism, and Other Difficult Positions.* New York: New York University Press, 2002.
Dickens, Charles. *The Pickwick Papers.* Clinton: Colonial Press, 1868.
———. *The Mudfog Papers.* New York: J. W. Lovell Company, 1883.
———. *Nicholas Nickleby.* 2 vols. London: Scholar Press, 1982.
———. *The Old Curiosity Shop.* London: Chapman & Hall, 1841.
———. *Barnaby Rudge.* Oxford: Oxford University Press, 2003.
———. *Martin Chuzzlewit.* 4 vols. New York: Hurd and Houghton, 1869.
———. *Dombey and Son.* London: Penguin, 2002.
———. *David Copperfield.* London: Penguin, 2004.
———. *Hard Times.* London: Penguin, 2003.
———. *Our Mutual Friend.* London: Everyman, 2000.
———. *The Selected Letters of Charles Dickens.* Edited by Jenny Hartley. Oxford: Oxford University Press, 2012.
[Dickens, Charles]. "New Year's Day." *Household Words* 19, no. 458 (1859): 97–102.
"Different Uses of Wood in the Arts". *The London Journal, and Weekly Record of Literature, Science, and Art* 36, no. 916 (1862): 136.

Dodd, William. *A Narrative of the Experience and Sufferings of William Dodd, A Factory Cripple. Written by Himself.* In *Factory Lives: Four Nineteenth-Century Working-Class Autobiographies*, edited by James R. Simmons, Jr., 181–300. London: Broadview, 2007.

Doyle, Arthur Conan. "The Man with the Twisted Lip." *Strand Magazine: An Illustrated Monthly* 2 (1891): 623–37.

Edgerton, David. *The Shock of the Old: Technology and Global History since 1900.* London: Profile, 2008.

Evans, Caroline. *Fashion at the Edge: Spectacle, Modernity and Deathliness.* London: Yale University Press, 2003.

Frank, Lawrence. *Charles Dickens and the Romantic Self.* Lincoln: University of Nebraska Press, 1984.

Gavin, Adrienne E. "Dickens, Wegg, and Wooden Legs." *Our Mutual Friend: The Scholarly Pages.* Accessed June 18, 2018. http://omf.ucsc.edu/london-1865/victorian-city/wooden-legs.html.

Gray, Frederick. *Automatic Mechanism, as Applied in the Construction of Artificial Limbs, in Cases of Amputation.* London: H. Renshaw, 1855.

Green, Harvey. *Wood: Craft, Culture, History.* London: Penguin, 2007.

Hall, Donald E., ed. *Muscular Christianity: Embodying the Victorian Age.* Cambridge: Cambridge University Press, 1994.

Hawthorne, Nathaniel. "Drowne's Wooden Image." In *Mosses from an Old Manse*, by Nathaniel Hawthorne, vol. 2, 75–91. Boston: Ticknor and Fields, 1854.

Herschbach, Lisa. "Prosthetic Reconstruction: Making the Industry, Re-Making the Body, Modelling the Nation." *History Workshop Journal* 44 (1997): 23–57.

Hingston, Kylee-Anne. "'Skins to jump into': The Slipperiness of Identity and the Body in Wilkie Collins's *No Name.*" *Victorian Literature and Culture* 40, no. 1 (2012): 117–35.

Holmes, Martha Stoddard. *Fictions of Affliction: Physical Disability in Victorian Culture.* Ann Arbor: University of Michigan Press, 2004.

Holmes, Oliver Wendell. "The Human Wheel, Its Spokes and Felloes." In *Sounds from the Atlantic*, by Oliver Wendell Holmes, 282–327. Boston: Ticknor and Fields, 1864.

Hood, Thomas. *Miss Kilmansegg and Her Precious Leg.* In *Selected Poems of Thomas Hood*, edited by John Clubbe, 193–270. Cambridge: Harvard University Press, 1970.

[Horne, Richard H.] "Dust; or Ugliness Redeemed." *Household Words* 1, no. 16 (1850): 379–84.

[Hudson, Thomas]. "The Cork Leg." Broadside Ballads Online. Accessed June 21, 2018. http://ballads.bodleian.ox.ac.uk/view/edition/5413.

Hugo, Victor. *Notre-Dame de Paris.* Trans. Alban Krailsheimer. Oxford: Oxford University Press, 2009.

Hutter, Albert D. "Dismemberment and Articulation in Our Mutual Friend." *Dickens Studies Annual* 11 (1983): 135–75.

[Jerrold, William Blanchard]. "Eyes Made to Order." *Household Words* 4, no. 81 (1851): 64–66.

Kafer, Alison. *Feminist, Queer, Crip*. Bloomington: Indiana University Press, 2013.

Ketabgian, Tamara. *The Lives of Machines: The Industrial Imaginary in Victorian Literature and Culture*. Ann Arbor: University of Michigan Press, 2011.

Kingsley, Charles. "The Science of Health." In *Sanitary and Social Lectures and Essays*, by Charles Kingsley, 21–40. n.p.: ReadHowYouWant, 2008.

LaCom, Cindy. "'The Time Is Sick and Out of Joint': Physical Disability in Victorian England." *PMLA* 120, no. 2 (2005): 547–52.

"Legs: Wooden and Otherwise". *All the Year Round* 14, no. 350 (1875): 463–67.

Lewes, George Henry. "Realism and the Art of the Novel." In *Literary Criticism of George Henry Lewes*, edited by Alice R. Kaminsky, 94–105. Lincoln: University of Nebraska Press, 1964.

Marchbanks, Paul. "From Caricature to Character: The Intellectually Disabled in Dickens's Novels." *Dickens Quarterly* 23 (2006): 3–14; 67–85; 169–81.

Marks, George E. *Marks' Patent Artificial Limbs with Rubber Hands and Feet*. New York: A. A. Marks, 1888.

Marx, Karl. *Capital: A Critical Analysis of Capitalist Production*. Translated by Samuel Moore and Edward Aveling. London: Kimble & Bradford, 1938.

Mayhew, Henry. *London Labour and the London Poor: A Cyclopaedia of the Condition and Earnings of Those That Will Work, Those That Cannot Work, and Those That Will Not Work*. Vol. 1. London: Griffin, Bohn, and Company, 1861.

———. *London Labour and the London Poor: A Cyclopaedia of the Condition and Earnings of Those That Will Work, Those That Cannot Work, and Those That Will Not Work*. Vol. 4. London: Griffin, Bohn, and Company, 1862.

Metzler, Irina. *A Social History of Disability in the Middle Ages: Cultural Considerations of Physical Impairment*. London: Routledge, 2013.

Mihm, Stephen. "'A Limb Which Shall Be Presentable in Polite Society': Prosthetic Technologies in the Nineteenth Century." In *Artificial Parts, Practical Lives: Modern Histories of Prosthetics*, edited by Katherine Ott, David Serlin, and Stephen Mihm, 282–99. New York: New York University Press, 2002.

Morgentaler, Goldie. "Dickens and the Scattered Identity of Silas Wegg." *Dickens Quarterly* 22, no. 2 (2005): n.pag.

[Morley, Henry]. "Ground in the Mill." *Household Words* 9, no. 213 (1854): 224–27.

"Mr. Wegg and His Class". *All the Year Round* 18, no. 441 (1877): 250–53.

O'Connor, Erin. "'Fractions of Men': Engendering Amputation in Victorian Culture." *Comparative Studies in Society and History* 39, no. 4 (1997): 742–77.

———. *Raw Material: Producing Pathology in Victorian Culture*. Durham: Duke University Press, 2000.

Oliver, Michael. *The Politics of Disablement.* Basingstoke: Macmillan Education, 1990.

Ott, Katherine. "Hard Wear and Soft Tissue: Craft and Commerce in Artificial Eyes." In *Artificial Parts, Practical Lives: Modern Histories of Prosthetics*, edited by Katherine Ott, David Serlin, and Stephen Mihm, 147–70. New York: New York University Press, 2002.

Owen, David. *English Philanthropy, 1660–1960.* Cambridge: Belknap Press of Harvard University Press, 1964.

Padilla, Jalynn Olsen. "Army of "Cripples": Northern Civil War Amputees, Disability, and Manhood in Victorian America." PhD diss., University of Delaware, 2007.

[Parker, Frances], the Countess of Morley. *The Flying Burgermaster: A Legend of the Black Forest.* n.p.: F. Morley, 1832.

Pendleton, John. "A Great Railway." *Good Words*, January 1898, 470–77.

Penner, Louise. "Dickens, Metropolitan Philosophy and the London Hospitals." *Victorian Medicine and Popular Culture*, edited by Louise Penner and Tabitha Sparks, 27–40. London: Routledge, 2015.

"Princely Travel in America". *All the Year Round* 8, no. 184 (1862): 174–180.

Reeve, Robin M. *The Industrial Revolution, 1750–1850.* London: University of London Press, 1971.

Ruskin, John. *Modern Painters.* Vol. 5. London: George Allen, 1906.

Russell, Marta. *Beyond Ramps: Disability at the End of the Social Contract.* Monroe: Common Courage Press, 1998.

[Sala, George A]. "The Great Hotel Question, in Three Chapters; Chapter the Third." *Household Words* 13, no. 310 (1856): 148–54.

Schweik, Susan M. *The Ugly Laws: Disability in Public.* New York: NYU Press, 2009.

Smiles, Samuel. *Self-Help; with Illustrations of Character and Conduct.* London: John Murray, 1859.

Smith, Charles Manby. "An Essay on Wooden Legs, with Some Account of Herr Von Holtzbein." *Tait's Edinburgh Magazine* 21, no. 244 (1854): 230–34.

———. *Curiosities of London Life; or, Phases, Physiological and Social, of the Great Metropolis.* London: W. and F. G. Cash, 1857.

[Smith, Sarah]. "The Lucky Leg." *Household Words* 19, no. 469 (1859): 374–80.

"The Mill Boy and the Hook". *The Spokane Press*, November 18, 1910.

"Street Sights in Constantinople". *All the Year Round*, no. 38 (1860): 279–84.

Sussman, Herbert, and Gerhard Joseph. "Prefiguring the Posthuman: Dickens and Prosthesis." *Victorian Literature and Culture* 32, no. 2 (2004): 617–28.

Sweet, Ryan. "Legs of Wegg and Others." Dickens *Our Mutual Friend* Reading Project. Last modified, August 5, 2015. https://dickensourmutualfriend.wordpress.com/2015/08/05/month-16-august-1865-legs-of-wegg-and-others/.

Taylor, Sunaura. *Beasts of Burden: Animal and Disability Liberation*. London: The New Press, 2017.
Van Ghent, Dorothy. *The English Novel: Form and Function*. New York: Harper and Row, 1953.
V. R. "The Wooden Legs in Dickens." *Notes and Queries* 171 (1936): 74–77.
Warne, Vanessa. "'To Invest a Cripple with Peculiar Interest': Artificial Legs and Upper-Class Amputees at Mid-Century." *Victorian Review* 35, no. 2 (2009): 83–100.
Williams-Searle, John. "Courting Risk: Disability, Masculinity, and Liability on Iowa's Railroads, 1868–1900." *The Annals of Iowa* 58, no. 1 (1999): 27–77.
Woloch, Alex. *The One vs. the Many: Minor Characters and the Space of the Protagonist in the Novel*. Princeton: Princeton University Press, 2003.

CHAPTER 5

"Losing a Leg to Gain a Wife": Marriage, Gender, and the Prosthetic Body Part

False rumps, false teeth, false hair, false faces,
Alas! poor man! how hard thy case is;
Instead of WOMAN—heavenly woman's charms,
To clasp cork, gum, wool, whalebone in his arms.
—"Made-up Beauty," *The Botanico-Medical Recorder* 8, no. 2 (1839): 30.

These lines from "Made-up Beauty," a poem that was first printed in the *New York Atlas* and then in the *Botanico-Medical Recorder* and the *Graham Journal of Health* (all in 1839), brings to the fore one of the common ways that gender inflected representations of prostheses in nineteenth-century Britain and America. In this example, men are portrayed as victims within a world increasingly filled with "false" females—women made up with various forms of cosmetic prostheses. This poem, on the one hand, draws from wider concerns about prostheses as devices that fraudulently mask physical aberrances—features that the physiognomically minded thought, if read correctly, could reveal character and behavioural traits. On the other, "Made-up Beauty" exposes specific concerns about the popularity of certain kinds of artificial body parts among women. These forms of artifice, which include dentures, wigs, and "false rumps" (a padded cushion-like accessory that was worn by women under their dresses) as well as makeup, are portrayed as robbing men of "woman's charms," by

R. Sweet, *Prosthetic Body Parts in Nineteenth-Century Literature and Culture*, Palgrave Studies in Nineteenth-Century Writing and Culture, https://doi.org/10.1007/978-3-030-78589-5_5

which the speaker appears to mean organic idiosyncrasies. The comment "how hard thy case is" has a double meaning, referring to both the supposedly unfortunate situation faced by men and the lamentable physical hardness or otherwise unhuman nature of some of the materials that were used to make prostheses: "cork," which many wrongly thought was used to make false limbs; "gum," which was used to make false teeth; wool, which was used to make false rumps; and "whalebone," which was used to make corsets and some high-end artificial legs.[1] The gendered focus of this poem, which looks at women from a male perspective, thus shows how certain types of prosthesis were associated with particular genders (cosmetic devices with female users) and how negatively these items could be received when detected. But how proscribed were stereotypes regarding prosthesis users? How were male and female prosthesis users represented in literary works that centred on romantic relationships between men and women? What do these representations tell us about attitudes to physical wholeness and difference and how they were inflected by gender?

This chapter traces representations of male and female prosthesis users in the marriage plot, the nineteenth-century narrative form most heavily populated by users of prosthetic devices. Building on the work of scholars such as Martha Stoddard Holmes and Talia Schaffer, who have brought attention to propensity for disabled figures to appear in Victorian marriage plots (a trope that has been labelled by Schaffer as "disability marriage"), I identify the prosthesis-marriage plot as a related yet separately identifiable formulaic narrative structure.[2] As I argue, when viewed collectively, and at times also individually, prosthesis-marriage plots presented unstable affective and imaginative treatments of prosthesis users. These representations shed light on the complex ways in which discourses of gender, class, and ableism intersected and how, in particular instances, the bodily status quo was challenged, brought into question, or even outright rejected. There were certainly fixed and distinct ideas associated with male and female prosthesis users. Nonetheless, prosthesis-marriage plots sometimes exploited and at other times interrogated these dominant attitudes.

The present chapter adds further nuance to the rich body of scholarship that centres on the Victorian marriage plot. From the seminal studies of Ian Watt and Lawrence Stone—which position representations of modern loving marriages as direct results of the rise of affective individualism in the seventeenth century—through Nancy Armstrong's ground-breaking work (which argues that eighteenth- and nineteenth-century marriage plots translated political tensions into personal desire) to more recent and revisionist appraisals, it is safe to say that the richness of the marriage plot as a literary mode has been matched by the vast amount of scholarship that has covered it.[3] But what has research in this field had to say about the prevalence of physical difference in nineteenth-century marriage plots? What do disability marriage plots tell us about contemporary attitudes to physical difference?

These questions have been partially addressed in recent studies by Cindy LaCom, Martha Stoddard Holmes, and Talia Schaffer, though fictional prosthesis users remain under-researched.[4] LaCom, Holmes, and Schaffer draw attention to the varied and profound ways the Victorian marriage plot grappled with the topic of disability. LaCom's 1997 essay "Is It More than Lame" explores the different attitudes and stereotypes associated with disabled women in terms of sexuality and motherhood. She argues that by reading Victorian writings about disabled women in both a literal and a metaphorical way, we can gain critical insights into not only gendered attitudes towards disability, but also an understanding of how woman's "nature" and "passion" have been historically constructed. Holmes builds on this work, focusing on the "melodramatization" of disability in the marriage plot.[5] She argues that melodramatic disability marriage plots of Edward Bulwer-Lytton, Charles Dickens, Dinah Mulock Craik, and Charlotte M. Yonge served a double function: these narratives "work[ed] through nondisabled women's desire, ... imagin[ing] the happy realms of able-bodied love by warning of the miseries that lay outside it"; they also "introduce[ed] and attempt[ed] to 'normalize' a number of potentially startling notions about disabled women."[6] Holmes, however, also reveals how Wilkie Collins positioned the disabled heroines of his sensation-fiction novels *Hide and Seek* (1854) and *Poor Miss Finch* (1872) as "direct challenges to melodramatic modes of representing disability."[7] Collins's disabled heroines transgressively express sexual desire and are, in the case of Lucilla Finch, procreative, in a manner that counters nineteenth-century marital norms.

Schaffer's more recent argument encompasses disability marriage but also refigures the way that we think about the Victorian marriage plot more broadly. Schaffer identifies the "familiar marriage" in the Victorian novel as a prevalent yet overlooked marital model that competed with romantic marriage, the form that has dominated the historiography of marriage heretofore. Schaffer explains that familiar marriage developed out of the eighteenth-century ideal of marrying for rational esteem rather than for romantic love, but, unlike the older model, it was predicated by existing affection. Familiar marriage, as it was imagined in literature, provided individuals an organic community and domestic influence, life choices that romantic marriage often failed to offer.[8] Regarding disability, Schaffer argues:

> [T]he disabled subject of the nineteenth century was the center of a social network. Because a disabled person required carers, this person was normally surrounded by others: parents, friends, servants, nurses. For a lonely person, a disabled partner could be the entry into a ready-made world, offering the intimate community ties for which so many Victorians yearned.[9]

Focusing primarily on the choices of nondisabled females, Schaffer explains that marriages with certain disabled male characters were attractive because of the social opportunities that they presented.

My discussion draws attention to the many instances where prosthesis users appear central to and sometimes disrupt the formulaic workings of the marriage plot. As I show, reflecting its status as a contested bodily device, the prosthesis was an unstable motif, at times facilitating either familiar or romantic marriage but at other times blockading the prospect of nuptials. Here, I begin by contextualizing gendered attitudes to physical difference and prosthesis use before parsing the various ways in which prosthesis users were represented in Victorian marriage plots. The literary investigation starts with depictions of prosthesis-using females and males whose prostheses are represented as obstacles to marital relations before investigating the various ways in which both male and female prosthesis users were imagined to be attractive. The sections that follow concentrate in turn on the sub-tropes of "love which conquers all reversals and disabilities," on devices that are alluring for economic reasons, on matches involving two prosthesis users, and on narratives that reject concealment and show prosthesis users as attractive on their own terms.

Gendered Difference

Before exploring gendered representations of prosthesis users in the marriage plot, it is first worth considering the different aspects that were at stake for men and women who lost body parts in the Victorian period. Although an appearance of physical wholeness was key for both men and women in this period, bodily losses were perceived to come at different costs for men and women. On the one hand, an impression of physical completeness was vital for men as it signalled moral integrity and social status; provided an index for youth and vitality (particularly in the case of teeth and hair); and distinguished them as individuals capable of working and thus accruing capital.[10] Women, in particular single ones, on the other hand, were under pressure to look as physically whole as possible, for the most part, in order to either look eligible for marriage or so as to not place the reputations of their husbands and families into disrepute by displaying physical incompleteness. Social class, however, was a factor that complicated clear distinctions between the social protocols for male and female prosthesis users. Functionality rather than aesthetics tended to be a priority for both male and female prosthesis users at the lower end of the social ladder as the work that they relied upon to make a living usually involved hard physical labour. But since their employment depended on positive relations with middle- or upper-class employers, and because social aspirations encouraged many to copy the behaviours and attitudes of their social superiors, the working classes were not exempt from bourgeois bodily discourses.

In terms of how men were affected by attitudes to physical difference in relation to the marriage market, we learn from writings such as *A Narrative of the Experience and Sufferings of William Dodd, a Factory Cripple* (1841) that the contemporary ableist masculine discourse of work also negatively impacted physically incomplete men's chances of marrying.[11] Though primarily concerned with drawing to public view the atrocious working and living conditions that apprentice factory children were exposed to, Dodd's memoir also shed light on contemporary attitudes to physical difference. Notably, Dodd describes the resistance that he encountered from a couple of women whom he attempted to court. The first "was too wise to join her destinies with those of a factory cripple. She left the town, and refused to answer my letters, which was a sufficient reason for my discontinuing to write."[12] The second, a housekeeper, was initially open to his advances but he soon realized that she was not serious about marrying:

> She would walk with me to church, to a place of amusement, to her relations to take tea, in the field, or anywhere but to the trap that I had baited for her. So I began to think that old birds were not to be caught with chaff. However, I did not like the idea of giving up to be laughed at, so I persevered, and pressed my suit more warmly, but soon found that she was only playing with me, like a cat with a mouse.[13]

Following this rejection, Dodd decided to give up on the hope of finding a partner and resigned himself to a life a bachelorhood. Though he was not a prosthesis user as such (owing most likely to the prohibitive cost and lack of functional use afforded by contemporary artificial arms), we learn about the difficulties that faced working-class amputee men wishing to marry. Key to the rejections that Dodd faced—in courtship as well as in his attempts to find work after his accident—was the assumption that, because he was missing a limb, he therefore lacked the necessary physical prowess to work to an acceptable standard and make a reasonable living. As indicated by his admission that his first lover "was too wise to join her destinies with those of a factory cripple" and his earlier claim that "to have married a factory girl, would only have involved both myself and her in greater troubles," Dodd viewed his own marital potential as diminished, underscoring the pervasive, insidious, and intersecting nature of ableist, masculinist, and capitalist discourses in this period.[14]

Men higher up the social ladder were also challenged by ableist attitudes associated with physical difference, though their chances of success were aided to some extent by their access to prosthetic body parts that could supposedly enable them to pass as normal. As Stephen Mihm notes, "[i]n an age of appearances, members of the middle classes necessarily hid their deformities and weaknesses, for fear that first impressions might deny them opportunities in marriage, employment, and social advancement."[15] Responding to demands for concealment, the prosthesis industry in Britain and America saw an unprecedented expansion, producing a litany of devices that prioritized enabling users to appear whole. Commenting on the success of high-end American limb prostheses, while emphasizing the necessity for respectable men to hide their physical difference in the private as well as public sphere, in his 1864 essay "The Human Wheel, Its Spokes and Felloes," the American physician and poet Oliver Wendell Holmes famously called for "[limbs] which shall be presentable in polite society." Holmes explains that "misfortunes of a certain obtrusiveness may be pitied, but are never tolerated under the chandeliers."[16] While the focus of Holmes's comments is on limb amputees, his comments on the intolerability of "misfortunes of a certain obtrusiveness" clearly extend to other forms of physical difference, including the perceived absence of other body parts including eyes, teeth, and even hair. Highlighting the extent to which eye loss was castigated as unattractive within the middle-class home, in 1862 Parisian ocular prosthesis maker Auguste Boissonneau remarked that living with eye loss was particularly hard to bear for men "on account of the moral torture occasioned by the humiliations or self-imposed idea of repulsiveness to which the unfortunate person who has experienced such a loss is exposed."[17] Though marriage per se was not a direct focus of Boissonneau's comments, a lexis of desire infiltrates his comments: he uses the term "repulsiveness." His words thus draw attention to the view that disablist attitudes to ocular difference prevented men with one eye from securing partners in respectable society.

Social responses to physical difference often provided obstacles to women seeking marriage partners in this period. Much of the stigma that negatively affected women drew from fears of contagious and hereditary risk. Drawing from contentious medical debates surrounding contagion, suggestiveness, and transmission, Martha Stoddard Holmes explains that, according to the logic of the time, "any physical impairment had the potential to be perceived as transmissible by contact; by miasmatic air; by a combination of contact, environment, and individual constitution; or perhaps simply by the social class into which one was born."[18] Hereditary fears about the implications of having physically aberrant mothers often ruled women who had lost body parts out of marrying. Numerous cultural and medical sources from the 1830s onwards perpetuated degenerative fears about all manner of disabilities being hereditary conditions. The perceived risks of allowing the disabled to copulate were made explicit by prominent medical experts such as Henry Maudsley:

> Certain unfavourable conditions of life tend unquestionably to produce degeneracy of the individual; the morbid predisposition so generated is thus transmitted to the next generation, and, if the unfavourable conditions continue, is aggravated in it; and thus is formed a morbid variety of the human kind, which is incapable of being a link in the line of progress of humanity.[19]

Such proto-eugenicist verdicts buttressed the troubling view that those exhibiting physical difference were not suitable marriage partners.

The fact that women, rather than men, were the primary focus of discussions of the supposed degenerative potential of disabled procreation brings to the fore a sexual double standard. In his study *Intermarriage* (1838), Alexander Walker stresses the sexual agency of men and focused attention on the need for women's bodies to be perfect: "the organization of the woman destined to reproduce, should be of the best kind; and that maturity, exercise and perfection in every function, are equally essential; for, as are these and their adaption to the male, so will be the perfection of the progeny."[20] While Walker identifies the necessity for both the male and the female partner to be physically sound in order to create perfect progeny, two factors reveal a gendered discrepancy. First, the medicalized (male) focus of this piece is on the female body (the male body was very much a second thought here). Secondly, both the authorship and the intended audience of this piece—both male—point towards the comparative agency that men had when it came to choosing their partners compared to women. Women's freedom of choice was restricted by social, economic, and even legal factors.[21]

As numerous critics have noted, during this period women's bodies were a constant source of obsessive cultural and medical attention.[22] Underlining the importance of physical beauty for women in this period, by the turn of the century, medical men, including Carl Heinrich Stratz and Havelock Ellis, were writing in detail about what constituted the perfect female body.[23] Advertisements for certain kinds of prostheses, such as dentures, were directly targeted at women, exploiting their anxieties to look physically whole and thereby presentable (see Fig. 5.1). Thus, while physical appearance was also important for men—more so, as we have seen, for those of higher social standing—for women it was paramount across the classes since without it their pathway to marriage, the standard route to social and financial security in adult life, was obstructed.

A 1909 *New York Times* letter to the editor entitled "Damages for an Eye" reveals in regrettable tones the effect that the loss of just a single eye could have to an aspiring young woman even by the early twentieth century.[24] Written by the aunt and guardian of a girl who lost an eye in a "street car collision," the piece argued that $4000—a figure equivalent to the "real wealth" value of $116,000 in 2019—which is the amount of compensation that the aunt's lawyers encourage her to aim for, was nowhere near enough to make up for the financial, social, educational, and potential medical implications of the niece's loss[25]:

> I think the accident has cut off her chances of ever marrying. No man would have her. Moreover, she has lost her judgment of distances, which will make her always clumsy in whatever occupation she may take up to support herself. On her blind side she is exposed to future accident, which she cannot be alert to guard against .... Her education is limited; the accident has handicapped her in acquiring more education. Her disfigurement has crushed her spirit—she feels ashamed, unfit to compete with people in the world.[26]

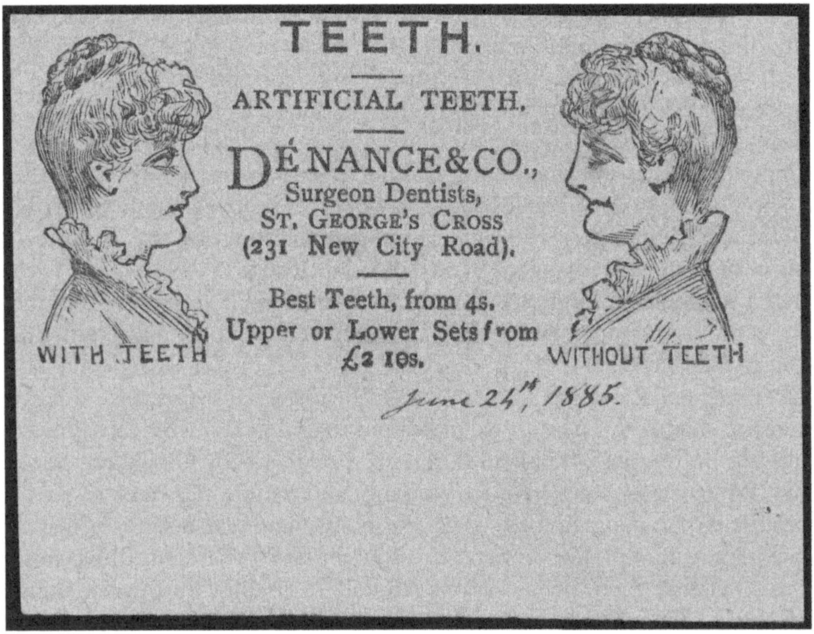

Fig. 5.1 An 1880s artificial teeth advertisement, depicting what a woman looks like with and without teeth. In the image without, her mouth looks noticeably sunken. Advertisement for Dénance & Co. artificial teeth, 1885, Wellcome Collection, London. Image courtesy of the Wellcome Collection. https://wellcomecollection.org/works/wnvnhxje. CC BY 4.0

From this passage, we can see the perceived price of physical loss for a young woman. The fact that the niece's compromised marriageability was the aunt's first consideration reveals the different issues at stake for men and women who experienced serious physical injury. This distinction is compounded further by the shame that the victim felt, an emotional state that stemmed not solely from the disfigurement itself but also from the unaccepting response of her contemporary society—since she no longer fitted the beautiful, physically complete, feminine ideal and thus faced marginalization.

We can contrast these marital fears with the financially linked but somewhat different anxieties felt by men who had lost functional parts. In the *Household Words* short story "The Lame Landlord's Story" (1867), a former engine driver recalls his initial fears after "losing" a leg and badly injuring an arm in what was thought an accident (but, as we learn, was actually an attempted murder): "It stood to reason that I couldn't hope to drive, nor even stoke, engines any more, and it really seemed that as if I'd nothing but the workhouse or a street-crossing before me." Here, the amputee's fear is not that he will be unable to marry—though this is later a short-lived anxiety that is quashed when his beloved "sent word that she'd love [him] better, without legs or arms, than any one else."[27] Instead, his immediate anxiety is that he will not only lose his job but also be rendered unemployable. As we learn from sources such as Dodd's memoir, for working-class men there was a clear link between productive potential and marital success, a reality somewhat obscured by the sentimental tone of "The Lame Landlord's Story." Nevertheless, what sources like this one and the *New York Times* article indicate is that men and women who were missing body parts often held different, though clearly entangled, priorities. While it was paramount for working- and middle-class men to pass as productive (to maintain their social, economic, and sexual status as masculine breadwinners), for women of equivalent social standings, the priority was to pass aesthetically as normal (to deflect disablist-misogynist stigma and afford them the best possible chance of marrying advantageously).

However, the concept that prostheses could enable physically incomplete men and women to appear whole to eligible partners (including not just those who had had limbs amputated, but also those who lacked hair, were missing teeth, or who had lost eyes) proved a point of contention. William Blanchard Jerrold famously debated the virtues and vices of prostheses by identifying that some observers saw artificial body parts as "emblem[s] of deceit," "device[s] of ingenious vanity," or items that "cover[ed] the wearer with gross and unpardonable deceit."[28] Drawing from such claims, an 1861 *Chambers's Journal* fictional work, which purports to be a factual article, suggests that women, especially ones looking for partners, had less of a right than men to use prostheses:

> I admit, if the lady I pay my addresses to has the misfortune to have one of her legs made of cork, I should prefer to be apprised of the fact before I put up the bans, rather than after the marriage-ceremony. Perhaps she, too, has some claim to be made acquainted with the circumstance, that my prepossessingly natural appearance is not altogether free from a certain alloy of unreality. But we will let that pass.[29]

This quotation draws our attention to the seemingly impossible position faced by a single woman who had lost a body part. While for her own sake, and her family's, a woman seen as incomplete would have likely faced significant pressure to use a prosthesis to mask her loss and thus hopefully attract a future husband, sources like this one imply that women should not try to deceive potential suitors. Yet this assertion is troubling when one considers the stigma that accompanied physical loss in this period. Men were encouraged to avoid copulating with physically aberrant women. Thus, women who were seen as deficient faced a conundrum: Did they use a prosthesis and risk discovery and subsequent accusations of fraudulence? Or did they give up on the possibility of marriage altogether and face an also stigmatized life of spinsterhood? Physically different men faced a similar problem but were less reliant on marrying for social and financial security, if they were deemed fit for work. As these situations reveal, life-shaping decisions about whether to use a prosthesis were at times informed less by impairment itself than by the social conditions that prohibited "intermarriage" with physically aberrant women and excluded non-normative bodies from the public sphere of work.

Prostheses in the Marriage Plot

"False" Females

As one can imagine, romantic relationships often do not proceed well for prosthesis users in nineteenth-century literary works. Many characters do not make it as far as engagement or even courtship. Pre-courtship narratives (i.e. short fictions centring on the early meetings of single subjects looking for prospective marital partners, such as those imagined in "Her Fatal Sneeze" [1890] and "The Stricken Fawn" [1893]), present prostheses as undesirable appendages that block potential paths to marriage.[30] Prostheses are depicted as turnoffs, which if inopportunely revealed foreclose romantic attraction. In "Her Fatal Sneeze," which (though not illustrated) was published in the halfpenny comic *Illustrated Chips*, a seemingly attractive "young thing of about twenty-four" has a mishap at a party that transforms her in the minds of the men present from "the belle of the evening" to a woman who must be avoided at all costs. Initially, the young woman, whose name is withheld, appears remarkably attractive: she has "rich, red lips, bright, sparkling eyes, and pearly teeth." Her teeth in particular "cause her to be the centre of attraction" as they "occasionally showed themselves as she smiled during her animated conversation." However, her façade of a pristine image is ruptured when she sneezes. To the "frightful" shock of her pursuers, her teeth—which we learn are false—fall out and her eye, which we realize is made of glass, "popped half-way across the room." Somewhat cruelly, the narrator reports the result of these unfortunately timed bodily acts: "Somehow after that the men did not seem to appreciate her beauty, and she was left severely alone."[31]

This unsympathetic treatment is symptomatic of both the popularity of cruel bodily humour, which followed the tradition set out by English jest-books in the mid-eighteenth century,[32] and wider prejudiced attitudes towards women, for whom physical blemishes were judged extremely harshly. Because of its low cost and accessible humour, *Illustrated Chips* was popular with working-class readers, at its height having a circulation of half a million.[33] The inclusion of satirized prosthesis users such as the false teeth user in this story shows how the failings of artificial body parts were considered humorous to working-class readers as well as those higher up the social ladder. What may have been particularly amusing to lower-class readers was the representation of a respectable woman's false teeth—most likely partial plates, which remained popular despite the advent of crowns and bridges—malfunctioning and causing a scene. Considering the still relatively (though no longer prohibitively) high commodity cost of artificial teeth in Britain in the 1890s—approximately 2 s. per tooth or 10–11d. for a full set—we might consider "Her Fatal Sneeze" as an example of the masses mocking the codes and practices of the more affluent.[34] In this regard, the sketch exposes the culture of using sophisticated, expensive prostheses to pass as a kind of folly, one always prone to catastrophe due to the social pressures involved: if one needs to use a prosthesis to pass as normal, then an unfortunate exposure will inevitably result in scandal.

Fun's "The Stricken Fawn," written for a more refined and affluent adult metropolitan middle-class readership, presents an even more hyperbolized story that emphasizes (and mocks) the shallowness of middle-class men, while utilizing the female prosthesis user for satirical purposes. Again, in this narrative, which positions itself as a parody of sentimental fiction, prostheses are obstacles to romantic relations, though here this representational model is also satirized. Set in the fictional seaside village "Drivelton-on-Sea," "The Stricken Fawn" pitches two sisters, Dolly and Dotty, against one another as potential wives for the absurdly named "Marquis of Puddlepond." The Marquis has the choice of Dolly, who is "weirdly beautiful" but compromised in terms of intellect (she stands at the window "catching flies") or Dotty, who is "intelligent" and "industrious," but "plain" due to her use of "the glass eye, false front, cork leg &c., &c." Faced with this choice, the Marquis selects the more physically conventional of the two sisters, Dolly, much to the disappointment of Dotty, who dies of heartbreak as a result.[35] In a plot that comically exaggerates the formulaic features of popular narratives that engage disability, "The Stricken Fawn" blatantly mocks the sentimental mode, while using prostheses as a familiar comic trope. Holmes identifies how the "twin structure," which "pairs a disabled woman with a nondisabled one and gives them distinctly different physical, emotional, and marital futures" was a common melodramatic trope in Victorian marriage plots.[36] This motif was clearly one of the satirical targets of "The Stricken Fawn."

A second obvious target was the image of the prosthesis as romantic obstacle, which is exposed as a related, hackneyed trope. The narrative treats both the "twin" and the prosthesis-as-romantic-obstacle structures with comic exaggeration. Dotty is not merely a false-teeth user but the user of many prosthetic devices—too many, it would seem, to be listed in entirety, as signified by the repetition of "&c." In this regard, we once again find evidence of ableism's instability in depictions of the prosthetic: though the writing is somewhat disablist in its use of physical (and, as I will show, mental) difference for humour, comic scorn is directed towards the superficiality of privileging physical beauty, an ideal shown to rely on a conspicuously whole body. Certainly, by Victorian standards of female beauty, Dolly is the more physically attractive of the two sisters, but society's over-investment in appearances is mocked through her diminished cognitive capacity. Ableism, in this case a favouring of normative intellect, paradoxically destabilizes preferences for physical normalcy. The story thus raises the ethically loaded question: What is preferable, a cognitively impaired or physically incomplete wife? In this case, though we are encouraged to question his decision, the Marquis decides that cognitive impairment is preferable, drawing attention to, while also satirizing, the social ideal of the married woman as silent and passive.

In other nineteenth-century narratives, prostheses prove obstacles further down the path to marriage. For instance, "Too Hard upon My Aunt" (1863), A. M.'s "Was She False" (1875), and "Kitty the Careless" (1883)—published respectively in *All the Year Round*, *The London Reader*, and *Judy*—use humour to communicate to diverse readerships advice regarding which prostheses women should avoid when looking for marital partners.[37] A further notable work in which a prosthetic device becomes an obstacle to marriage is Thomas Hardy's *The Woodlanders*, which was first published serially in *Macmillan's Magazine* and *Harper's Bazar* from 1886 to 1887.[38] In this "woodland story," which as Penny Boumelha observes, "giv[es] narrative form and textual representation to desire,"[39] Felice Charmond (one of the novel's more unlikeable sexual transgressors) is eventually revealed to her adulterous lover, the also odious philosopher-physician Giles Fitzpiers, as a wig user. This revelation has disastrous consequences for their illicit affair.

Unlike other plots in which prosthesis use is withheld from the reader as well as the non-prostheticized party in the romance, in Hardy's novel dramatic irony is present throughout.[40] As early as the second chapter, we witness the aptly named master barber Mr. Percomb, who we learn is employed by the affluent widow Mrs. Charmond, as he attempts to convince the rural peasant Marty South to part with her hair so that a wig can be made for his client. Part of the rhetorical strategy utilized by Percomb (whose name splices "peruke" and "hair comb") relies on the revelation of his client's identity: we learn that Mrs. Charmond has taken a fancy to Marty's abundant chestnut curls and that she is also, conveniently, Marty's father's landlady. Before the reader directly encounters Mrs. Charmond, our impression of her is already tarnished by the strong-arm tactics employed to procure Marty's hair and by the apparent reason for her wanting to attain a wig. As Marty explains herself: "She wants my curls to get another lover with; though if stories are true she's broke the heart of many a noble gentleman already."[41] Once the transaction is complete, which occurs after Marty realizes that her own romantic ambitions involving the apple picker Giles Winterborne are futile, she is sworn to secrecy regarding the identity of the new user of her hair. But this pact does not stop her from revealing this news much later in the narrative. After she catches wind of Mrs. Charmond's affair with the married Fitzpiers, she writes to the latter to inform him that his mistress wears a wig. The narrator notes: "It was poor Marty's only card, and she played it, knowing nothing of fashion, and thinking her revelation a fatal one for a lover."[42] Though Fitzpiers is slow to read Marty's correspondence, its effect is the one anticipated by the peasant girl (despite the narrator's scepticism): the physician satirically jeers Charmond following the revelation, resulting in their breakup and Fitzpiers's "abrupt departure" from their elopement on the Continent. However, the implications of this end to Charmond and Fitzpiers's relations have even graver consequences for the former than

Marty, or indeed anyone, could have possibly expected. Dramatic justice is served brutally, since Charmond, in her attempts to follow Fitzpiers and make amends, is confronted by a former South Carolinian lover and then shot dead after an impassioned argument.

The punishment that Charmond faces for her misdeeds in *The Woodlanders* is certainly harsher than those facing other discovered prosthesis users in similar plots. Always present in plots in which female prosthesis users are revealed and subsequently rejected is the suggestion that to wear false body parts is indicative of a predilection for fraudulence more generally—a point that supports Jerrold's observations. Implicitly, women who mask their physical differences by using prosthetic body parts might be concealing more than simply non-normative anatomy. They might, for example, be fallen women in disguise, as in the case of Isabelle Vane in Ellen Wood's *East Lynne* (1860–1861), who is able to masquerade as a governess in her estranged husband's family home by employing the use of blue-tinted spectacles and a veil.[43] By comparison, Louisa May Alcott's similarly sensational 1866 novella *Behind a Mask* depicts an actress who uses false hair and teeth to pose as a governess and eventually marry her way into an aristocratic position.[44]

Female prosthesis users were thus in many cases akin to or sometimes literally actresses, who, as Tracy C. Davis notes, remained associated with prostitution and inappropriate sexual conduct.[45] The link between false body parts and sexual misdirection was certainly present, for instance, in A. M.'s "Was She False?": a short story whose very name implies deception beyond mere material means. Here, Uncle Mortimer's response to discovering that his wife-to-be wears false teeth in this tale is revelatory. Upon discovery, he exclaims: "She's treacherous! I have been deceived all through. I daresay the rest of her is as false as her—but no matter! I am disenchanted at last. I have bidden her an eternal adieu!"[46] To Mortimer, as signified by his inference that "the rest of her" is false and as his inability to complete his sentence, the prosthetic part is a synecdoche of falsity more generally. This view is supported in the 1892 *Pick-Me-Up* comic article "Women: By Our Office Boy," which (in cockney dialect) critiques what it sees as a fashion for fakery among women: "Look how a woman 'fakes' herself! Paint, powder, false teeth, false hare, and—oh, that I should have to rite it!—a false buzzum, wich, by-the-by, gen'rally covers a false hart!"[47] The same link between what we might call primarily cosmetic prostheses and sexual deviance was forged in the so-called "Pepper and Salt" section of *Judy* in 1886. A short sketch narrates the consequences after a jealous wife discovers another woman's set of false teeth in her husband's pocket. Such a disclosure once again associates female prosthesis use with sexual misconduct.[48]

Responding to rapid developments in prosthetic technologies, the somewhat embellished reportage that this progress received, and the apparent ubiquity of certain forms of prostheses—for example, wigs and dentures—some narratives even imagined scenarios where only after marriage did partners realize that they had wedded prosthesis users. One of the most infamous instances where this scenario plays out occurs in another of Hardy major works, *Jude the Obscure* (1895).[49] In this brutal tale of crushed aspirations, one of the main impediments facing the eponymous protagonist in his quest for educational, spiritual, and familial success is the fatal attraction that he develops (and can never quite quash) towards Arabella Donn, the devious daughter of a pig farmer. We learn relatively early on, after Jude and Arabella's hasty nuptials, that one of several layers of deception that Arabella uses to ensnare her husband is an appearance of abundant hair produced by prosthesis use. On the very evening of their marriage, to his disappointment, Jude discovers that Arabella uses chignons (these are false curls of the kind represented visually in Fig. 5.2):

> A little chill overspread him at her first unrobing. A long tail of hair, which Arabella wore twisted up in an enormous knob at the back of her head, was deliberately unfastened, stroked out, and hung upon the looking glass which he had bought her.
>
> "What—it wasn't your own?" he said, with a sudden distaste for her.[50]

Jude's reaction—shock, "distaste," and incredulity—is telling of late nineteenth-century responses to false hair, as well as more general standards of women's beauty, which preferred a natural look. But Arabella's response to Jude brings to the fore how social expectations ironically also informed her decision to use prosthetic hair additions. Reacting to Jude's question as to whether the hair is not her own, Arabella explains: "O no—it never is nowadays with the better class." To this comment, the clearly riled Jude responds: "Nonsense! Perhaps not in towns. But in the country it is supposed to be different. Besides, you've enough of your own, surely?" But once again, Arabella uses convention to defend herself: "Yes, enough as country notions go. But in towns the men expect more .... The more you have the better in Aldbrickham .... Every lady of position wears false hair—the barber's assistant told me so."[51] Here, Arabella aligns her

Fig. 5.2 A magazine plate depicting various elaborate chignons, individually and in use on attractive, youthful female users. "New coiffures for morning and evening toilettes," plate 109 to *Myra's Journal of Dress and Fashion*, February 1, 1888, wood engraving, 26.2 × 18.6 cm, Wellcome Collection, London. Image courtesy of the Wellcome Collection. https://wellcomecollection.org/works/e8bxupkm. CC BY 4.0

decision to use chignons with metropolitan middle-class tastes, at once revealing Jude's ignorance of supposedly refined urban sentiments—a factor contributing to his eventual failure as a scholar in Christminister—and asserting her provocative forwardness in dealing with the opposite sex. Her frankness ironically contrasts the effect of her prosthesis use, which she admits to deploying in order to deceitfully attract male suitors.

False hair is also clearly used as synecdoche for Arabella's character in *Jude the Obscure*: the false part is representative of a fraudulent character. She elsewhere sucks in her cheeks to give an impression of dimples; lies about pregnancy to secure Jude's hand in marriage; and plies him with alcohol, preying on his physical and spiritual weakness, to force him to remarry her. Her final and most repugnant duplicitous act comes after she neglects her duties as Jude's nurse while he is battling severe illness. Returning to her lodgings to check on him, she discovers that he has died but pretends that he is asleep so that she can continue to participate in the festivities of Christminister's "Remembrance games." Though not a disability, the way that Arabella's physical difference (her artificial-hair use) functions as a signifier for character flaws—in this case excessive self-interest, fraudulence, and a lack of empathy. Arabella's false hair is thus certainly treated negatively—matching the stigma attached to female prosthesis users discussed earlier in this chapter—but at the same time its use is decidedly savvy and in line with metropolitan middle-class bodily expectations, which ironically favoured natural beauty while encouraging women to use hair additions.

Key to Hardy's exploration of wigs and the factors that give rise to their use is the role of the prosthesis maker. Note how the authority that Arabella leans on in her justification of wig use is a "barber's assistant."[52] Though only referred to in passing by Arabella, the inclusion of this figure's statements as an explanation for why chignons are needed underlines the influence that these professionals had on consumers. Arabella's commitment to wig-use and her reference to a barber's assistant to support her case shows her susceptibility to marketing, implying that other aspirational women from her social station might also be vulnerable to these tactics. Contemporary wig makers and barbers, not unlike other kinds of prosthesis makers, used a variety of techniques to entice consumers. British wig makers, including Thomas Elliott and "Professor Brown" used literary components in their advertisements to promise female users "restored youthfulness, beauty and the ability to go unnoticed as 'incomplete' women."[53] With, what is fair to say, a clear degree of scorn, Hardy's work implies that such commercial rhetoric was effective among consumers of Arabella's social standing. By using dark humour and elements of the grotesque as representational modes, *Jude the Obscure* suggests that the trickery underpinning the modern women's ability to assume a perception of beauty when she is in fact by contemporary standards incomplete is an issue that cannot be entirely blamed on the female consumers of prostheses. Assistant barbers, and by extension other professionals associated with the prosthesis trade, are exposed as responsible for fuelling trends that encourage the use of artifice.

Although socially conscious writers such as Hardy were cautious not to pin the blame entirely on women for what was perceived as an insidious growth of artifice as the nineteenth century progressed, it is clear from the examples given thus far that a disproportionate stress in the prosthesis-marriage plot was placed on women as supposedly duplicitous consumers of artificial body parts. On the one hand, we might consider the prevalence of female prosthesis users in marriage plots as a symptom of what Margrit Shildrick explains as the positioning of women "*vis-à-vis* an inaccessible body ideal," one which she also notes is even more difficult to attain for disabled women.[54] For postmodern feminists such as Shildrick and Sandra Lee Bartky, who utilize a Foucauldian approach to women's bodies, we can consider norms of behaviour that affect women's bodies as examples of "disciplinary power"[55]: "The reiteration of the technologies of power speak to a body that remains always in a state of pre-resolution, whose boundaries are never secured. Indeed, repetition indicates its own necessary failure to establish any stable body, let along an ideal one."[56] Continual references to female prosthesis users whose supposedly flawed bodies ultimately stand in the way of happy marriages highlights on the one hand the disciplinary pressures that women faced to conform to normative bodily standards. But the repetition of the trope also speaks to a failure to establish bodily order amid a society that not only demanded physically normative women but also did not fully tolerate the use of devices designed to facilitate an impression of such.

By further emphasizing the instability of prostheticizing as a disciplinary act for incomplete women, female prosthesis users, however, were not always entirely negatively construed in romantic contexts—even when they were packaged in a discriminatory manner. For example, the racing journalist Joe Capp's humorous poem "Rewination!" describes a situation in which a man entices his wife to bed only to discover that she uses a range of prostheses:

"Oh, come to my arms," said he, said he;
"Oh come to my arms," said he.
And she came to his arms with all her charms,
False hair, false teeth, false calves, false arms,
Yes, she came to his arms, came she.

"Now, let us retire," said he, said he;
"Now let us retire," said he.
Then she placed her false teeth in the water so clear
Unstripped her false arm, and took off an ear.
Lay back in her chair as if for a rest,
Began once again by removing her chest,
While her husband in horror looked on.

"I see you have sold me, old lady," said he.
"I see you have sold me," he said.
"But please tell me now, you deceitful old she,
How much of you is coming to bed,
How much of you is coming to bed."[57]

The husband in this poem looks on "in horror" during the undressing scene and addresses his wife as a "deceitful old she," conforming to the views of earlier Victorians. The poem also clearly draws from similar scenes depicted in earlier literary writings, such as Jonathan Swift's 1734 poem "A Beautiful Young Nymph Going to Bed."[58] The title "Rewination!" suggests that the husband has been ruined by the discovery of his partner's artificiality. But the poem is not entirely stable in its treatment of prostheses. In the final two lines, we see how little the husband is deterred. He asks jestingly—perhaps even flirtatiously—how much of his wife is coming to bed, but he does not flee or retract his proposition. In fact, he repeats it twice, emphasizing his hardly wavering sexual desire. In this context, the prosthesis user is stigmatized but ultimately not diminished sexually or cast off socially. As I show later, other Victorian narratives imagined both male and female prosthesis users as not only tolerable but sometimes desirable in romantic relationships.

Marriageable Men?

Though female users tended to be the focus of prosthesis narratives of failed marriages, some did represent male prosthesis users whose relationships end due to the response of their partners to their physical difference.

For instance, the 1893 *Hearth and Home* article "Good Advice and a Wooden Leg," which was initialled "G. W. C.," used an anecdote about a man with a wooden leg to make a point about the virtues of giving out advice. In this somewhat bizarre piece, the writer, who we assume is male, tells a story about a consumptive friend of his whom he advised to go and live on the slopes of the Rocky Mountains, "in the pine woods, in a hut or tent," until "dead or well." The writer describes how he bumped into the same friend several years later, only to be shunned by him after the friend blamed him for the loss of his leg, which we learn was injured by a grizzly bear. This injury, compounded with the fitting of a wooden leg, had dire consequences for the once-consumptive man: though cured of the disease, his fiancée broke off their engagement as she "don't want a husband with a wooden leg."[59] What is curious about this case is the blame that is attached to the wooden leg rather than the limb loss itself. On the one hand, we might understand the "wooden leg" as functioning as a linguistic supplement for a "missing leg," with its linguistic usage paralleling its use in "real life" as a substitute for an absent part. On the other hand, we might take this phrasing to imply that there is something particularly off-putting about a wooden leg, perhaps suggesting its unsettling uncanniness or uncouth physical appearance. A key factor to consider is the readership for this piece, which we see in the subtitle: *An Illustrated Weekly Journal for Gentlewomen*. The price per issue is also relevant: 3d. The affluent women that this publication was aimed at would have no doubt agreed with the requirements described by Oliver Wendell Holmes for limbs "which shall be presentable in polite society." What is inferred by the description "wooden leg" is a slight on the primitive design of the chosen device and what such a choice supposedly reveals about the affluence, social standing, and/or taste of its user. If the device is not a limb "which shall be presentable in polite society," then to a gentlewomen reader neither is its user.[60]

We learn from another fin-de-siècle short story, André de Blaumont's "My Fiancé's Glass Eye," that related prejudices surrounded male glass-eye users. Published in Alfred Ludlow White's *Short Stories: A Magazine of Select Fiction* (1894) and then in the Massachusetts pro-Catholic newspaper *The Sacred Heart Review* under the variant spelling "Andree de Beaumont" in 1895, the light-hearted domestic narrative tells the story of an engagement that is almost broken off after the wife-to-be is mistakenly led to believe that her betrothed is a glass-eye user.[61] Written from the perspective of the man's fiancée, Angela, early on the story describes her potential husband, Raoul, in glowing terms. The opening lines read: "He was tall and dark, to my eyes charming in every respect." Angela's devotion to Raoul is exemplified as she decides to reveal to him all of her "faults" as, being an honest woman, she feels that it is best if they are known before they are married—a tactic that clearly contrasts with those of several of the prosthesis users described so far. Following this revelation, Angela asks what Raoul's faults are, at which point he becomes embarrassed and says nothing. We then learn that he is short-sighted and uses a monocle. Soon after this discussion, Raoul departs to get his affairs in order so that he can marry his beloved. Shortly after his departure, the housemaid, Justine, brings Angela and her mother the shocking news that Raoul has left behind his "glass eye." The responses of Angela and her mother are telling: Angela initially bursts into tears: "Mamma, mamma! He has a glass eye, it is frightful! I shall never get over it! I shall die of mortification!" Her mother, with an air of self-assurance, consoles her: "This gentleman has deceived us, that is all. To tell the truth, I always found something strange in his looks." Angela's mortification soon turns to tenderness, however, once she imagines that Raoul most likely lost his eye "in some honourable, noble way" and feels that she still loves him "even with his one eye." Her mother's stance is somewhat more hard-line. Responding to her daughter's quick change of heart, she asks,

> Are you crazy? Can you imagine for a moment that I would permit you to marry a man with such an infirmity? As pretty as you are, and only nineteen! No, a thousand time no, my child, so do not create a romance of devotion and sacrifice, for it is useless. I would never consent to your marrying a one-eyed man. Why, if he should lose the other eye he would be blind! How pleasant that would be![62]

Fortunately, the dispute does not escalate further, despite Angela's assertion that she "would be his faithful dog, to lead him about and protect him"—a statement that indicates the respect with which the canine companions of the vision-impaired were sometimes treated in the Victorian imagination.[63] When Justine presents Raoul's "glass eye" to Angela, it becomes apparent that the house servant confused the man's monocle with an ocular prosthesis.[64]

Regarding the use of prostheses in marriage plots, this short story reveals several relevant points. First, the narrative makes apparent the instability of attitudes towards male prosthesis users in terms of marital potential. The conflicting views that Angela and her mother express suggest that there is more to the issue than simply the view that using a prosthetic device disqualified a man from the marriage market. There is certainly a considerable level of revulsion expressed towards the prospect of Raoul being a glass-eye user, and, like female prosthesis users, he is accused of deceit. But Angela is also quick to forgive him, guessing that he injured his eye performing a noble deed. Though Raoul is affected by some of the same stigma that the female prosthesis users above suffer, more leeway is provided to the assumed male prosthesis user. While little to no attention is paid to the causes of prosthesis use for female users in the stories above, leaving it to the Victorian reader's imagination to assume what was seen as the worst—that their physical losses are congenital and thereby potentially heritable—the narrator's concern with this issue implies that there were preferable causes for physical losses among men. These causes could mitigate concerns regarding marrying a physically incomplete man. As David Serlin has observed, since the American Civil War prostheses and other visual markings that proved injury became "visual shorthand *for* military service." "Disability," Serlin remarks, "became the permanent uniform worn by those who participated in the aftermath of civil warfare."[65] Literary works, such as Charlotte Brontë's *Jane Eyre* (1847), also depict injuries including limb loss and blindness as potential markers of noble or redemptive deeds. Rochester sustains these injuries while attempting to rescue his estranged wife from the house fire that she started.[66] A gendered double standard for prosthesis use is brought to the fore by "My Fiancé's Glass Eye": respect or honour might be hastily bestowed on some men with acquired body losses, maintaining their status as eligible bachelors, but women with similar physical differences were rarely treated in this way.

There is a further aspect of Justine's misrecognition of a prosthetic eye. On the one hand, we can say that de Blaumont/Beaumont's narrative uses female working-class ignorance as a narrative and comic prop: Justine's confusion of a monocle with a glass eye drives the story and brings comic relief once her mistake has been made apparent. But on the other hand, in an illuminating manner, the story draws our attention to the understandable lack of knowledge that working-class women might have regarding sophisticated prosthetic technologies. Justine's confusion certainly brings to the fore semantic questions. What constitutes a prosthesis? Is a monocle not in fact a kind of glass eye? But de Blaumont/Beaumont's story also hints at the inaccessibility of prostheses to the contemporary working classes. Artificial eyes were sold to surgeons for $10 in New York-based John Reynders & Co.'s 1889 *Illustrated Catalogue and Price List of Surgical Instruments*—a price relative to the "income value" of $2890/£2258 in 2019.[67] Charles Lenz & Sons of Philadelphia offered a rather improved price of $5 circa 1892, but this still represented a high commodity price—a price relative to the "income value" of $1300/£1016 in 2019.[68] In 1893, Charles Traux, Greene & Co offered artificial eyes from the "Finest French and German Manufacturers" for the trade price of $3.50 each or directly to patients for $10 each—prices relative to the "income value" of $989/£773 and $2830/£2211 respectively in 2019.[69] On the other side of the Atlantic, the Down Bros. of London sold artificial eyes made to order for £1 1 s. in their 1890 catalogue—relative to the "income value" of £817/$1046 in 2019.[70] Given the average hourly compensation for unskilled work in America was $0.14 in 1894, and that in the same year average annual nominal earnings in Britain were just short of £62,[71] one can begin to see from an economic standpoint why a servant might not have been familiar with these devices if their master or mistress did not use them: when one considers the added profit margin of the surgeon or ophthalmologist selling the device on top of all but one of the prices provided above, we can conclude that purchasing an artificial eye would have involved a considerable outlay for someone in Justine's position.

Monocles, meanwhile, were, as Richard Corson notes, often used as "a matter of fashion in imitation of the aristocracy."[72] Marius Hentea observes how ubiquitous these devices were among the intelligentsia during the period of high modernism in the early twentieth century.[73] House servants, such as Justine, it is clear, both economically and socially speaking, fell far outside the target social market for both monocles and glass eyes, in part making Justine's ignorance in this narrative an entirely plausible representation. One also should consider the fact that if you had good eyesight and were not missing an eye—as seems to be the case with Justine—there might be very little reason for one to know the differences between eyeglasses and glass eyes, especially given the pressing concerns facing individuals in her social position, including stark economic realities and long working hours.

"Love Which Conquers All Reversals and Disabilities"

The scandal in "My Fiancé's Glass Eye" in the end proves to be much ado about nothing since the misrecognition of the monocle as a glass eye becomes clear. But regarding gender, the story implies that if a gentleman's social position might not be adversely affected by his use of an ocular prosthesis—in other words, if it might be "presentable in polite society" and not a barrier to work—then it might be tolerated by a female suitor. We also learn that prosthesis use might be accepted romantically speaking, if the injuries leading to its use were sustained performing noble deeds. In the way that it engaged with themes of devotion and tolerance, de Blaumont/Beaumont's story draws from the sentimental mode, which often deploys the motif of "love which conquers all reversals and disabilities."[74] Earlier in the century, on the other side of Atlantic, Charles Dickens drew from and tested the limits of this trope in his 1840s novels *Barnaby Rudge* (1840–1841) and *Dombey and Son* (1846–1848).[75] In these works, Dickens depicts circumstances in which men are made more attractive by their physical differences and/or prosthesis use. In *Barnaby Rudge*, a novel set in the late eighteenth century, the son of an inn keeper, Joe Willet, only manages to secure the heart of the woman he has long admired, Dolly Varden, after he loses an arm fighting for the British against American revolutionaries. Indeed, though the once coquettish Varden rejects Joe's advances before his departure to America, his return injured stimulates the development of both warm sentiments and emotional maturity in Dolly.

Less conventionally, and more relevant to the literary history of prostheses, Captain Cuttle of Dickens's slightly later novel *Dombey and Son* is another amputee male who is depicted as romantically attractive. Though his prosthesis use (he has a hand with interchangeable heads) is not directly described or implied as an enticing feature, the pursuit of him by his landlady, the widow Mrs. MacStinger, a figure as fierce and intimidating as her surname would suggest, is one of the main comical threads in Dickens's novel. In this regard, the novel riffs off the prosthesis-marriage plot trope. Dickens uses an extended metaphor of naval battle to describe the somewhat odd and one-sided romantic subplot connecting Cuttle and MacStinger: the widow is "the enemy" from whom the Captain, with the help of his allies, attempts to avoid "discovery and capture."[76] Inverting the typical prosthesis-marriage plot, in which either a physically normative male mistakenly pursues a female prosthesis user or a prosthesis-using male hopelessly courts a bodily normative female, here a seemingly non-disabled female goes after a man whose prosthesis use is conspicuous. Cuttle's prosthesis is in fact a defining feature according to Dickens's narrator. Cuttle is first introduced as "a gentleman in a wide suit of blue, with a hook instead of a hand attached to his right wrist,"[77] and the device is in regular use throughout the novel. For example, he arranges his hair with it[78]; he replaces the hook with a knife attachment and peels with it[79]; and he exercises "great power" over young boys with it.[80] Though not present when Cuttle performs many of these actions—because he spends most of the novel hiding from her—MacStinger is not in the least deterred by the Captain's use of a replaceable hook hand. In fact, after he escapes his lodgings at her home—thanks to the assistance of his apprentice, Rob the Grinder—her pursuit of him, though we do not witness it as readers, is described as relentless. First, we learn that MacStinger would "never hear of his deserting" her and her children, leading Cuttle to the "desperate determination of running away." After his escape, Cuttle appears to be "too well acquainted with the determined and dauntless character of Mrs. MacStinger, to doubt that that heroic woman had devoted herself to the task of his discovery and capture."[81]

Cuttle's worst fears are actualized when, assisted by her children, MacStinger eventually finds and accosts him at the Midshipman, his new home, which he shares with the nautical instrument maker Solomon Gills.[82] Thankfully for Cuttle, he is rescued by his friend and fellow skipper Captain Bunsby, who consoles MacStinger, takes her home, and, as we later learn, is ensnared into marrying her. Curiously, like Cuttle, Bunsby displays physical difference: he is described as a "bulk-head—human, and very large—with one stationary eye in the mahogany face, and one revolving one, on the principle of some lighthouses."[83] Bunsby is slightly later described "to be always on the look-out for something in the extremest distance, and to have no ocular knowledge of anything within ten miles."[84] Certainly Bunsby is a good fit for the nautical theme that Dickens develops in the novel—his fixed eye resembles a telescope, always looking into the distance, and his revolving eye takes after a compass, moving and adjusting in chaotic scenarios to identify the right course—but we could also read him as a possible artificial-eye user due to the signifiers provided: one of his eyes is stationary and carries a permanent expression as though it is looking into the far distance—for instance, when Cuttle talks to him in Chapter 39, his eye "look[ed] fixedly at the coast of Greenland." These descriptions align with Jerrold's description of how artificial-eye users appeared prior to the advent of Auguste Boissonneau's enamel eyes, which first came to Britain in the early 1850s: "While one eye was gazing intently in your face, the other was fixed in another direction—immovable, the more hideous because at first you mistook it for a natural eye. A smile may overspread the face, animate the lip, and lighten up the natural eye; but there was the glass eye—fixed, lustreless, and dead."[85] Bunsby's depiction fits this description, though Dickens never explicitly states whether he is an ocular-prosthesis user. Either way, it is intriguing that MacStinger's affections are displaced from a hook-hand user onto a figure whose physical difference is also manifest. It is implied, therefore, that MacStinger is either not deterred by or perhaps even directly attracted to men whose bodies would have been considered contemporaneously as non-normative.

A key concern for MacStinger appears to be the care of her loving but somewhat unruly children, for whom Cuttle claims to have done "a world of good turns."[86] Her attraction to physically different men might thus be partially explained by her apparent belief in the potential of these men for excelling in the line of domestic work. Because they are neither youthful nor physically whole, they appear, by contemporary standards, no longer suitable for the labour market, therefore making them, in the eyes of MacStinger, available for work within the home. As implied by the "great power" that Cuttle is said to exercise over MacStinger's young boys with his hook, the physical differences of these Captains make them, for MacStinger, uniquely suited to childcare.[87] In this regard, we might consider this depiction as exemplifying Schaffer's argument about depictions of "familiar marriages" in Victorian literature: "instead of feeling erotic desire for another's body, familiar marriage advocates sympathetic helpfulness; instead of demanding a private dyad, familiar marriage wants a larger social network; instead of valorizing the self-made man, familiar marriage prizes unselfish participation in a system of mutual care."[88] In the case of *Dombey and Son*, we see that Cuttle and later Bunsby offer opportunities to MacStinger that will enlarge her social network and enhance the lives of her children in return for other forms of care—cooking, cleaning, and so on. Dickens, though, hardly sets up this form of marriage as an exemplary model. Both Cuttle and Bunsby fear MacStinger and they appear resistant to the kind of life that a union with her presents. They both describe such an arrangement in humorous terms as "capter."[89] Though I am resistant to interpreting Cuttle's prosthesis use as symbolic of "castration," as John Jordan suggests, it is clear that this language of imprisonment implies fear regarding the believed emasculation that figures like Cuttle and Bunsby might experience if faced with a life of domestic servitude.[90]

Enticing Devices

While Dickens experimented with the idea that amputees might appear attractive to women in certain social positions due to the potential that they offer in extending social networks, in other writings prostheses were represented as attractive because of the economic status that they signified. One narrative that imagined a prosthetic body part as an attractive feature on a male user was Geoffry Brandon's 1899 short story "The Story of the Jewel in the Stage Coach as Related by Simon Simple Barrister-at-Law."[91] The piece was published in the *Ludgate*, an illustrated family magazine, which was a less-successful competitor to the *Strand Magazine*—a title made popular in the 1890s through to the 1920s thanks largely to the success of Arthur Conan Doyle's Sherlock Holmes short stories. Brandon's narrative tells the story of a relationship between a daughter of a lawyer and a gentleman that begins with an odd first encounter: the spinster Mistress Kesia Clawby attempts to grab what she believes to be a jewel resting on the back seat of a dark London coach, only to find that the glimmering object is in fact the glass eye of a small man, who we later discover to be a judge, Sir Benjamin Cossett. Not knowing that Mistress Kesia is the woman who "assaulted" him on the coach, Sir Benjamin warms to the lawyer's daughter at a later soirée. The narrator notes that Mistress Kesia "was certainly not a 'giddy flirt.'" Witnessing this relationship kindling, the first-person narrator cynically notes: "I perceived what sort of card the good lady was playing, and that she meant to own that jewel of a glass eye after all." Mistress Kesia and Sir Benjamin go on to marry, emphasizing how little the former was deterred by the latter's use of a prosthesis, which is even said to look "fishy" following the poke administered on the coach.[92]

Brandon's story is not dissimilar to the *Cheshire Observer*'s 1879 comic sketch "He Fixed Her with His Glassy Eye," in which a man uses his glass eye as a prop to woo a woman whom he finds attractive. When asked why he is staring at her, he returns: "It's a glass eye madam—only a glass eye. But I'm not surprised that even a glass eye should feel interested in so pretty a woman."[93] Unlike this sketch, though, in which the prosthetic body part serves an advantageous rhetorical purpose, in Brandon's short story, rather than being an undesirable physical feature, Sir Benjamin's jewel-like glass eye is a signifier of wealth and prosperity. We can see from the relatively high contemporary cost of ocular prostheses stated earlier (£1 1 s. for medical practitioners in Britain in 1890) that one would need to be reasonably wealthy to consider purchasing such a device. Given that average annual nominal earnings were £56 10s. 1d. in 1890, we can surmise that ocular-prosthesis use was a fairly consistent indicator of prosperity.[94] In this social context, a high-end prosthesis, rather than detracting from a man's attractiveness, might signify wealth and social status, arguably the two most important qualities for men to exude within the contemporary marriage market.

As Vanessa Warne shows, however, female prosthesis users were also imagined as attractive partners for financial reasons in certain Victorian literary works. Warne draws our attention to Hood's poem *Miss Kilmansegg and Her Precious Leg* and Smith's short story "The Lucky Leg," arguing that "[b]oth texts ... deal with the logic of attraction: the marriage of an able-bodied man to a disabled woman involves either the acquisition of new wealth or the preservation of already inherited wealth."[95] At the centre of Hood's poem is a marriage plot involving the affluent eponymous protagonist who, following a riding accident that results in the amputation of her right leg, demands a prosthesis made of solid gold. As we have seen in Chap. 3, she is courted by and eventually marries an Italian Count, who later murders her using "Her Precious Leg" so that he can have it melted down and used to service his considerable gambling debts. Smith's slightly more positive depiction, on the other hand, tells the story of the working-class amputee Mary Wigley who is mysteriously courted and later married by an affluent man, named Mr. Gordon. As the mystery behind Gordon's "peculiar interest" is revealed, we learn that he had two previous wives, whom had both died. Both women were wooden-leg users, and the first left her estate to Gordon on the condition that he married another amputee. Warne is understandably more positive about Smith's story, but she cautions against reading either as particularly progressive. Regarding Hood's poem, she notes that it "complicates the conventional association of disability with dependency" and seems somewhat progressive in the way that Miss Kilmansegg chooses to use an obviously artificial device, resisting the mandate to pass as normal. However, as she goes on to explain, "It is ... difficult to read the poem in general as progressive. Miss Kilmansegg is unembarrassed by her disability, because her loss of a leg allows her to control a large portion of her family's wealth. It also provides her with a means of displaying this wealth."[96] Of course, also problematic is the way that Miss Kilmansegg is killed off as a kind of heavy-handed, moralistic punishment for her ostentation. By comparison, Smith's story is somewhat different in terms of class politics as it traces the social rise of an amputee, who is seemingly rewarded for acting virtuously as a female amputee. She keeps her physical difference discreet and even allows her guardian to write to Mr. Gordon to inform him of her disability after it is feared that he has overlooked the fact. But clearly this depiction is heavily influenced by conservative attitudes regarding physical difference. Though the story ends with the provocative question "who knows all of the advantages of disadvantages?" Smith depicts Wigley's successful fate as the result of a combination of rare favourable circumstances and behaviour underpinned by bodily normative and patriarchal values: Mary makes every attempt to conceal her difference, is honest when called upon, and is largely passive throughout the story.[97] As we will see later, however, other writings from this period imagined prosthesis users as attractive for less fortuitous and more radical reasons.

Prosthetic Matches

As I have demonstrated here, prosthesis-marriage plots were sometimes represented as primarily economic exchanges, a trope that was transgressive in the way that it resisted positioning those with physical differences as dependents but ultimately not entirely progressive in terms of depicting them as attractive in other ways. Another romantic plot structure that was radical in some ways but also tainted by conservatism was that which represented prosthesis users marrying each other. The results of such a scenario were depicted in the 1819 etching *Il faut des epoux assortis Dans les liens du Mariage* [*Persons in Wedlock Should Be Properly Matched*] (see Fig. 5.3), which shows a couple displaying the various prostheses that they use, including artificial buttocks, breasts, hair, eyes, and teeth. Similarly, in the *New Monthly Magazine/New York Monthly Magazine* M. Sullivan's serialized novel *Stronger than Death*, we see that "two ineligible candidates for marriage," a woman with a glass eye and a man with a wooden leg, are united in wedlock.[98]

A further prosthesis marriage plot, which was popular in both nineteenth-century Britain and America, which utilized a similar trope, was Heinrich Zschokke's short story "*Das Bein*"—first published in the

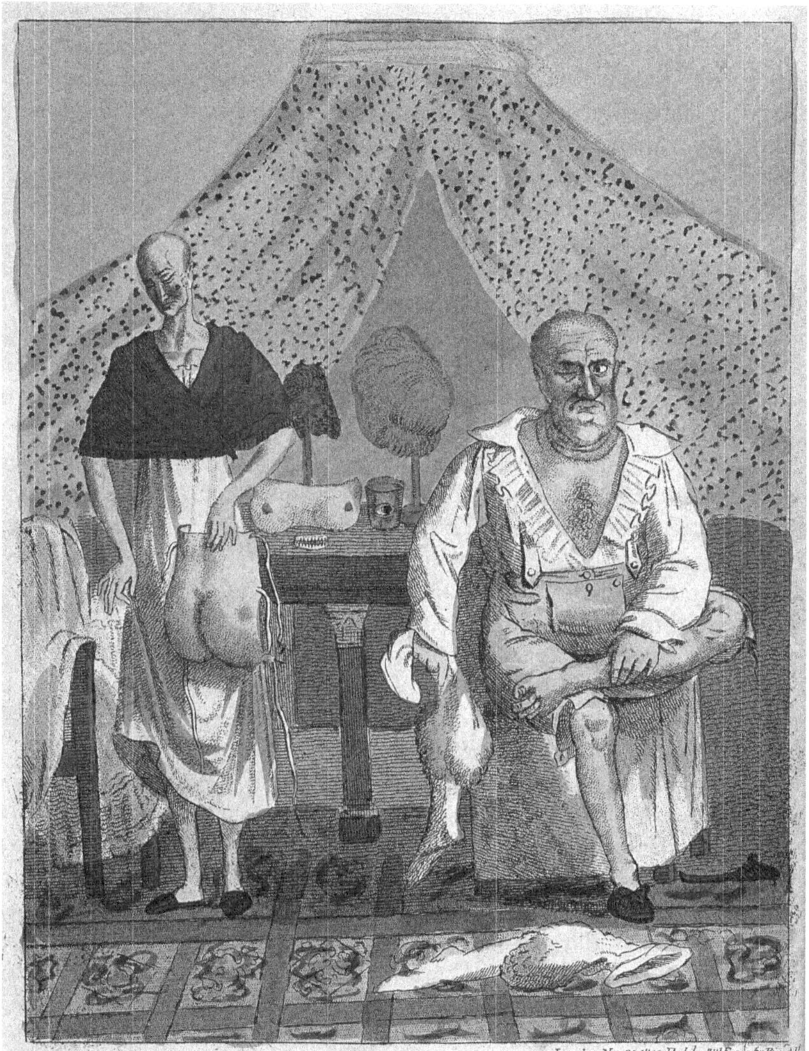

Fig. 5.3 An early nineteenth-century coloured etching that portrays a couple underdressing, removing their various prostheses, which include artificial buttocks, breasts, hair, eyes, and teeth. *Il faut des epoux assortis Dans les liens du Mariage*, 1819, etching, 25 × 19.2 cm, Wellcome Collection, London. Image courtesy of the Wellcome Collection. https://wellcomecollection.org/works/fwk3ub5b. CC BY 4.0

German monthly magazine *Erheiterungen* in 1811.[99] Zschokke's narrative appeared in Britain in the *Kaleidoscope* in 1827 and was subsequently reprinted—often without acknowledgement of its original author or translator—at least fifteen times in British newspapers and periodicals and twenty-nine times in the American press between 1827 and 1906.[100] These reprintings appeared under various titles, including "The Leg," "Was He a Madman?" "Losing a Leg to Gain a Wife," "Love's Amputation; or, A Leg for a Lady," "Giving a Limb for a Wife, with Future Regrets," and "A Case in Surgery. Which Shows That a Man May Do Too Much for Love of Woman."[101] Under its various guises, this somewhat misogynist tale narrates the French surgeon Lewis Thevenet's encounter with the Englishman Charles Temple. The latter travels to Calais, has the surgeon collected from his home by carriage, and then tries to convince him to amputate his right, uninjured and unimpaired, leg. At first, Thevenet outright rejects Temple's demand, maintaining his stance even when the latter points a pistol at him. But the surgeon later resigns himself to performing the operation to spare Temple additional suffering after the Englishman turns the gun on himself, threatening to shoot his own leg to necessitate the procedure. We later learn from a letter of thanks sent to Thevenet by Temple that the latter demanded the amputation so that he could marry the woman he loved, a fellow amputee who rejected his first marriage proposal when he had two legs. As Temple explains: "Miss Harley was a wonder of beauty, but she had but one leg. [O]n account of this imperfection she feared to become my wife lest I should esteem her the less for it." Here, the burden of social attitudes to physical difference initially proves an obstacle to romantic relations, though a second amputation becomes a catalyst. In the same letter, Temple recites how Miss Harley was the first person that he visited after returning to England as an amputee, and that the two were married shortly thereafter. He even entrusts to her the sacrifice that he made after marriage and he claims that "She loves me now the more affectionately." Still, as Thevenet predicts, Temple comes to regret his sacrifice. When the two men meet in England some years afterwards (Thevenet flees France during the Revolution), Temple reveals that his marriage did not work out. He explains: "her wooden leg prevented her dancing, so she betook herself to cards and to fashions. There is no such thing as living peaceably with her." Reflecting on his physical sacrifice, he tells Thevenet, "It was a silly adventure. Had I my leg again, I would not give the paring of a nail. Between you and me, I was a FOOL."[102]

The extent to which the narrative appears progressive in its treatment of amputees as marriageable is compromised by its conservative conclusion, which reinstates the physically whole body as sacrosanct. Though the attraction towards an amputee woman is shown to be so strong that a nondisabled male willingly has a leg amputated to better his chances of marrying her, such a sacrifice is shown to be foolish when we learn of the unhappy outcome of their matrimony. To use disability-studies scholar Michael Oliver's language, the story ultimately conforms to a view of disability as a "personal tragedy."[103] It provides historical evidence for Alison Kafer's observations about how ableism affects imaginings of disabled futurity: "If disability is conceptualized as a terrible unending tragedy, then any future that includes disability can only be a future to avoid."[104] The narrative also draws attention to the physical and social drawbacks of Temple's decision, further reinforcing the hegemony of wholeness: Temple can barely stand up when reunited with Thevenet and notes himself that his "cursed leg" is a "hindrance to [him] in everything"; he also observes that he would have "been an admiral of the blue [a squadron of the Royal Navy], had not [his] wooden leg disqualified [him] from the service of [his] country."[105] Here, though his wooden leg was represented initially as a catalyst to marriage, it is ultimately depicted as an obstacle, which not only brings him unhappiness and physical difficulties, but also emasculates him by removing him from the social arena of work.

Zschokke's conclusion aligns the stories about other cultural representations of prostheses, which satirize the sacrifices that certain women expected their partners to make for them. For instance, the wood engraving shown in Fig. 5.4 depicts a couple passing by the window display of an "*ocularise*" (glass-eye maker). Clearly responding to the popularity of the artificial eyes produced by Parisian prosthetist Auguste Boissonneau, who coined the term "*ocularise*," the woman tells her blue-eyed lover that the man of her dreams has black eyes and asks if he could obtain artificial eyes to remedy this discrepancy. Sources like this one show how, within romantic plots, prostheses were both the targets of and the tools used for misogynist satires involving women with marital aspirations.

Countering Concealment

While prosthesis users were most often framed as eligible in marriage plots owing to financial reasons, out of sympathy, because their prospective partners were also physically different, or because of the social prospects

Fig. 5.4 A mid-century wood engraving, which mocked the supposedly outrageous demands that young women made of their partners. Here a couple pass by an ocularist's window and the woman asks her partner whether he would consider changing his eyes to darker ones. *Yeux Artificiels*, c. 1868, wood engraving, Wellcome Collection, London. Image courtesy of the Wellcome Collection. https://wellcomecollection.org/works/eu4hegsx. CC BY 4.0

that a marriage with them offered, on rare occasions prosthesis use was framed as an attractive feature in its own right. Arguably, the most radical nineteenth-century prosthesis marriage plot was Robert Williams

Buchanan's 1862 *Temple Bar* short story "Lady Letitia's Lilliput Hand." In this final section, I show that Buchanan's narrative presents a prosthesis user whose transgressive marital success stems from her eventual decision to embrace rather than conceal her physical difference.

Buchanan's "Lady Letitia's Lilliput Hand" was written in the style of sensation fiction and published when that literary mode was at its height in 1862. It therefore appeared almost a decade before the author's career-defining public spat with Algernon Charles Swinburne and Dante Gabriel Rossetti, in which he infamously claimed that these Pre-Raphaelite poets belong to a "fleshly school of poetry"—a charge that seems somewhat rich after learning more about Buchanan's earlier short story.[106] "Lady Letitia's Lilliput Hand" explores the history of an alluring yet highly mysterious woman whose most attractive features are her conspicuously dainty, perfectly formed hands. As the narrator remarks, "it was by her beautiful hands that the Lady Letitia achieved her choicest triumphs. Hands so tinily, delicately lovely were never imitated by sculptor; and when she waved them before her slaves, the heart was hushed with admiration."[107] Helena Michie has identified how common it was for women's hands to function as sexual symbols in the Victorian novel.[108] As Clare Stainthorp puts it: "Nineteenth-century norms of female dress pushed the localisation of the erotic to the peripheries of the body, the hand being one such site."[109] But in the case of Lady's Letitia, the narrator's "normative positivism" (in this case, his idolization of the physically normative female form) is complicated once we learn more about one of her hands in particular.[110]

We first meet Lady Letitia amid a scramble of suitors attempting to wed her. The frontrunner for her affections, the wealthy artist Edward Vansittart, is kept away by the protagonist after a second mysterious character, Mr. Montague Vernon, appears on the scene. As we move forward, we learn that this new arrival is a disguised figure from Lady Letitia's past. Vernon is, in fact, Louis Carr, the once fraudulent associate of her late husband, the financially ruined gambling addict Lord Augustus Marlowe. The narrator reveals that Vernon/Carr had previously attempted to win Letitia's affections by manipulative and duplicitous means and intends to do so again by blackmailing her with misleading evidence that portrays her as Lord Augustus's murderer. The visual motif that Buchanan centres on when the unravelling mystery is Lady Letitia's wonderfully formed left hand, which we learn is prosthetic. The narrator reveals that her hand was amputated after it was crushed by a heavy window during a deathbed struggle with her late husband, who had poisoned himself to frame Lady Letitia as his murderer, a desperate attempt at revenge after discovering that Carr's attempts to woo his wife had made progress. All ends well, however. Lady Letitia manages to counter Vernon/Carr's blackmail by threatening to claim him as an accomplice and reveal his identity to the authorities—a move that would be disastrous for the blackmailer since he is wanted for gambling fraud. Carr flees but is unluckily arrested and later hanged for his crimes, while Lady Letitia and Vansittart are happily reunited. Running counter to contemporary marital norms, Vansittart marries Lady Letitia even after she reveals all, including the fact that her most prized asset, her hand, is artificial. As a kind of postscript for the narrative, the narrator includes a note from Vansittart commenting on his marriage with Lady Letitia. We learn that she was a great wife and that she died after giving birth to his eldest daughter. Underlining the narrative and symbolic work that the prosthetic body part does in this story, it ends with a Gothic yet sentimentalized image: Vansittart reveals that he keeps "the Lilliput hand" as "a memento."[111]

Buchanan's short story draws special attention to the sensory difference of Lady Letitia's prosthesis. The bodily variation that her hand presents is a source of mystery, intrigue, and grotesqueness. Her hand is cold to touch, providing a grotesque morbidity as well as a sense of foreboding and uncanniness to her literary depiction. The coldness of her hand provides a hint of her non-normativity, echoing the concern of the British prosthesis maker Henry Heather Bigg, who in 1855 lamented the fact that "touch instantly decides between the real hand and its counterfeit."[112] Adding to the grotesqueness of Buchanan's depiction, there is a shocking scene in which her hand is stabbed right the way through without causing so much as a drop of blood or cry of pain. Her hand is thus presented as conspicuously different to an organic one in that it both feels non-normative and is unable to feel itself. The that her hand is prosthetic exacerbates the tension between her passive, touched, aristocratic, feminine hand and the active, touching, middle-class, masculine hands of her suitors.[113] In aesthetic terms, the hand acts as a Gothic motif, a kind of uncanny vestige of the past that works as a sensational plot device. We could also label the stabbing scene as an instance of "bodily shock," a trademark of the sensation-fiction genre. The centrality of Lady Letitia's prosthetic hand to nexus of the plot provides evidence for Martha Stoddard Holmes's and Mark Mossman's argument that "[d]isability can be seen as central to the very poetics of sensation fiction."[114] However, there is clearly embedded in Buchanan's narrative a comment on the pressures of normativity. Though underpinned by a way of thinking about the body that privileges organic physical completeness, the haptic critique of the prosthetic that this story brings to the fore questions the efficacy and potential implications of using a hand that looks real but feels fake. "Lady Letitia's Lilliput Hand" suggests that using such a device will always eventually invite suspicion, a feeling that no prosthesis user from this period wished to evoke, since doing so would undermine her/his ability to pass.

The greatest resistance to "normative positivism" (the preference for normalcy) and "non-normative negativism" (disablism) that Buchanan's short story provides comes in the form of the non-normative marital success that Lady Lilliput eventually achieves.[115] Ultimately, she is able to recapture Vansittart's affection and marry him—not by concealing her use of a prosthesis but by revealing its artificiality and the scandalous back story that its use conceals. Kafer's concept of "crip time" provides an illuminating model for Buchanan's representation as not merely conforming to, but rather disrupting, ableist traditions. Kafer conceptualizes "crip time" in opposition to common curative trends in imagining disabled futurity, which see disability as something that should be avoided or cured at all costs. According to ableist thinking, the only conceivable alternative to being cured is a life of "unending tragedy." Kafer instead seeks to imagine "more accessible futures, ... [a] yearning for an elsewhere—and, perhaps, an 'elsewhen'—in which disability is understood otherwise: as political, as valuable, as integral."[116] In other words, Kafer's project is to challenge the ableist thinking in our present that affects the real and imagined futures of people living with disabilities. It would be remiss to say that "Lady Letitia's Lilliput Hand" participates in this kind of disruptive and progressive crip project, but what can be argued is that the disabled future that it imagines for the amputee Lady Lilliput is one that resists the hegemonic normative ways of thinking about disabled women in relation to marital futures. Lady Letitia can thus be understood in relation to the transgressive, bold, and sexually active disabled female characters of Wilkie Collins's sensation fiction—Madonna Blyth of *Hide and Seek* (1854) and Lucilla Finch of *Poor Miss Finch* (1872), most notably.[117] Unlike these novels, or indeed the "familiar" disabled marriage narratives described by Schaffer—which present disabled characters as integral parts of domestic marital family networks—here Buchanan's work provides a direct affront to the normative positivism engendered by the curative concept of the concealing prosthesis, since the narrative emphasizes the success and attraction of disavowing passing as the primary prerogative for a prosthesis user.[118] Vansittart's decision to keep hold of the Lilliput Hand after his wife's death is at once fetishistic and reveals a transgressive fondness for physical difference.

Clearly, many nineteenth-century writers were fascinated by the marital prospects facing prosthesis users, highlighting the growing influence of prosthesis culture within the Victorian imagination. But when considered together, there seems to be little consensus regarding how prosthesis users should be treated in literary writings. On the one hand, prostheses were often represented, in particular when used by women, as devices that might disqualify the user from the marriage market. But on the other, they were sometimes imagined as either tolerable or outright attractive for a wide variety of reasons: they could kindle sympathy and "love which conquers all reversals and disabilities"; they could signify wealth and social status or be economically valuable in their own right; they could be used to attract other prosthesis users; or they could themselves be aesthetically pleasing fetish objects whose artificiality is made manifest. As this chapter shows, such representations shed light on the complex and variegated ways that gender inflected attitudes to physical wholeness, difference, and the devices used to conceal it. Fictional marriage plots tended to focus on fitness, productive potential, and social status, sometimes providing more leeway to male prosthesis users compared to their female equivalents. Representations of female prosthesis users, meanwhile, often centred on tensions between the visibility and the invisibility of their devices and the perceived honesty of the user. Considered together, these sources provide evidence of how unstable literary attitudes were regarding the concept of physical wholeness. Plainly, ableism and patriarchalism were entangled influences in the period investigated, and yet both male and female prosthesis users remained conspicuously and ironically visible in the literary works of a society that proved increasingly interested in concealing such differences from public view.

Notes

1. American limb maker John S. Drake used whale ivory in the construction of his devices. See John S. Drake, *Drake's Patent Artificial Legs, Hands, Arms, &c.* (Boston: J. Drake, 1859). Also see Chapter 2, especially the discussion of Ahab's ivory leg in Herman Melville's *Moby-Dick* (1850).
2. Martha Stoddard Holmes, *Fictions of Affliction: Physical Disability in Victorian Culture* (Ann Arbor: University of Michigan Press, 2004); Talia Schaffer, *Romance's Rival: Familiar Marriage in Victorian Fiction* (Oxford: Oxford University Press, 2016).

3. Ian Watt, *The Rise of the Novel: Studies in Defoe, Richardson, and Fielding* (Berkeley: University of California Press, 1974); Lawrence Stone, *The Family, Sex and Marriage in England 1500–1800* (London: Penguin Books, 1990); Nancy Armstrong, *Desire and Domestic Fiction: A Political History of the Novel* (Oxford: Oxford University Press, 1987); Ruth Perry, *Novel Relations: The Transformation of Kinship in English Literature and Culture, 1748–1818* (Cambridge: Cambridge University Press, 2004); Helena Michie, *Victorian Honeymoons: Journeys to the Conjugal* (Cambridge: Cambridge University Press, 2006); Helena Michie, *Love among the Archives* (Edinburgh University Press, 2015); Sharon Marcus, *Between Women: Friendship, Desire, and Marriage in Victorian England* (Princeton: Princeton University Press, 2007); Elsie B. Michie, *The Vulgar Question of Money: Heiresses, Materialism, and the Novel of Manners from Jane Austen to Henry James* (Baltimore: Johns Hopkins University Press, 2011); Maia McAleavey, *The Bigamy Plot: Sensation and Convention in the Victorian Novel* (Cambridge: Cambridge University Press, 2015); Schaffer, *Romance's Rival.*
4. Cindy LaCom, "'Is It More than Lame': Female Disability, Sexuality, and the Maternal in the Nineteenth-Century Novel," in *The Body and Physical Difference: Discourses of Disability*, ed. Sharon L. Snyder and David T. Mitchell (Ann Arbor: University of Michigan Press, 1997); Holmes, *Fictions*; Schaffer, *Romance's Rival.*
5. Holmes, *Fictions*, 35.
6. Ibid., 72–73.
7. Ibid., 74.
8. Schaffer, *Romance's Rival*, 2–3.
9. Ibid., 160.
10. See Bruce Haley, *The Healthy Body and Victorian Culture* (Cambridge: Harvard University Press, 1978), 3–22; Erin O'Connor, *Raw Material: Producing Pathology in Victorian Culture* (Durham: Duke University Press, 2000), 104–106; Katherine Ott, "Hard Wear and Soft Tissue: Craft and Commerce in Artificial Eyes," in *Artificial Parts, Practical Lives: Modern Histories of Prosthetics*, eds. Katherine Ott, David Serlin, and Stephen Mihm (New York: New York University Press, 2002), 152–53; and Stephen Mihm, "'A Limb Which Shall Be Presentable in Polite Society': Prosthetic Technologies in the Nineteenth Century," in *Artificial Parts, Practical Lives: Modern Histories of Prosthetics*, eds. Katherine Ott, David Serlin, and Stephen Mihm (New York: New York University Press, 2002).
11. William Dodd, "A Narrative of the Experience and Sufferings of William Dodd, A Factory Cripple. Written by Himself," in *Factory Lives: Four Nineteenth-Century Working-Class Autobiographies*, ed. James R. Simmons Jr. (London: Broadview, 2007).

12. Ibid., 211.
13. Ibid.
14. Ibid., 201.
15. Mihm, "A Limb," 288.
16. Oliver Wendell Holmes, "The Human Wheel, Its Spokes and Felloes," in *Sounds from the Atlantic*, by Oliver Wendell Holmes (Boston: Ticknor and Fields, 1864), 306–307.
17. Auguste Boissonneau, *General Observations on Artificial Eyes, Their Adaption, Employment and the Means of Procuring Them* (Paris: J.-B. Baillière and Son, 1862), 5.
18. Holmes, *Fictions*, 63.
19. Henry Maudsley, "Galstonian Lecture II on Relations between Body and Mind, and between Mental and Other Disorders of Nervous System," *Lancet*, April 30, 1870, 609–10; qtd. in Holmes, *Fictions*, 66–68.
20. Alexander Walker, *Intermarriage; or the Mode in Which, and the Causes Why, Beauty, Health, and Intellect Result from Certain Unions, and Deformity, Disease and Insanity from Others* (London: John Churchill, 1838), 324.
21. See Shirley Foster, *Victorian Women's Fiction: Marriage, Freedom, and the Individual* (London: Routledge, 2012); Mary Lyndon Shanley, *Feminism, Marriage, and the Law in Victorian England* (Princeton: Princeton University Press, 1993); and Ben Griffin, *The Politics of Gender in Victorian Britain: Masculinity, Political Culture and the Struggle for Women's Rights* (Cambridge: Cambridge University Press, 2012).
22. See Helena Michie, *The Flesh Made Word: Female Figures and Women's Bodies* (Oxford: Oxford University Press, 1987); Mary Poovey, *Uneven Developments the Ideological Work of Gender in Mid-Victorian England* (Chicago: University of Chicago Press, 1988); Athena Vrettos, *Somatic Fictions: Imagining Illness in Victorian Culture* (Stanford: Stanford University Press, 1995); Pamela K. Gilbert, *Disease, Desire and the Body in Victorian Women's Popular Novels* (Cambridge: Cambridge University Press, 1997); Alison Bashford, *Purity and Pollution: Gender, Embodiment, and Victorian Medicine* (Basingstoke: Palgrave Macmillan, 1998); and Andrew Mangham and Greta Depledge, *The Female Body in Medicine and Literature* (Liverpool: Liverpool University Press, 2013).
23. Carl Heinrich Stratz, *Die Schönheit des weiblichen Körpers*, 14th ed. (Stuttgart: F. Enke, 1903); Havelock Ellis, *Man and Woman: A Study of Human Secondary Sexual Characters* (London: Walter Scott, 1904).
24. "Damages for an Eye: What Would Compensate for Mental Anguish and Disability?" *New York Times*, December 4, 1909.
25. Samuel H. Williamson, "Seven Ways to Compute the Relative Value of a U.S. Dollar Amount, 1774 to Present," Measuring Worth, accessed July 19, 2021, www.measuringworth.com/uscompare/.

26. "Damages for an Eye," 10.
27. "The Lame Landlord's Story," *Temple Bar* 20 (1867): 127.
28. William Blanchard Jerrold, "Eyes Made to Order," *Household Words* 4, no. 81 (1851): 64.
29. "The False Hair," *Chambers's Journal of Popular Literature, Science and Arts* 8, no. 396 (1861): [65]–67.
30. "Her Fatal Sneeze," *Illustrated Chips*, no. 2 (1890): 26; "The Stricken Fawn." *Fun* 58, no. 1489 (1893): 219.
31. "Her Fatal Sneeze," 26.
32. Simon Dickie, "Hilarity and Pitilessness in the Mid-Eighteenth Century: English Jestbook Humor," *Eighteenth-Century Studies* 37, no. 1 (2003); Simon Dickie, *Cruelty and Laughter: Forgotten Comic Literature and the Unsentimental Eighteenth Century* (Chicago: University of Chicago Press, 2011).
33. John Storey, *The Making of English Popular Culture* (London: Routledge, 2016).
34. Prices here are taken from the following sources: Holborn Dental Institute, "Advertisement for the Holborn Dental Institute," 1886, British Library Online Gallery, accessed July 5, 2018, http://www.bl.uk/onlinegallery/onlineex/evancoll/a/014eva000000000u05139000.html; Mr. Foley, Surgeon Dentist, "Advertisement for Mr. Foley, Surgeon Dentist," c.1896, Wellcome Collection, accessed July 5, 2018, https://wellcomecollection.org/works/j625k4vh.
35. "The Stricken Fawn."
36. Holmes, *Fictions*, 38.
37. "Too Hard upon My Aunt," *All the Year Round* 10, no. 242 (1863); A. M., "Was She False?" *The London Reader: Of Literature, Science, Art and General Information* 26, no. 656 (1875); "Kitty the Careless," *Judy*, August 1, 1883; Ryan Sweet, "'Get the Best Article in the Market': Prostheses for Women in Nineteenth-Century Literature and Commerce," in *Rethinking Modern Histories of Prostheses in Anglo-American Commodity Cultures, 1850–1960*, ed. Claire L. Jones (Manchester: Manchester University Press, 2017), 126.
38. Thomas Hardy, *The Woodlanders* (Oxford: Oxford University Press, 2009).
39. Penny Boumelha, "Introduction," in *The Woodlanders*, by Thomas Hardy (Oxford: Oxford University Press, 2009), xxvi.
40. Edgar Allan Poe, "The Spectacles," in *Prose Tales (Second Series) Arthur Gordon Pym*, by Edgar Allan Poe (Boston: Dana Estes, 1884).
41. Hardy, *The Woodlanders*, 13.
42. Ibid., 220.
43. Ellen Wood, *East Lynne* (Oxford: Oxford University Press, 2008).

44. Louisa May Alcott, *Behind a Mask: or, A Woman's Power*, in *Louisa May Alcott Unmasked: Collected Thrillers*, ed. Madeleine Stern (Boston: Northeastern University Press, 1995).
45. Tracy C. Davis, "Actresses and Prostitutes in Victorian London," *Theatre Research International* 13, no. 3 (1988): 221–34.
46. A. M., "Was She," 76.
47. "Women. By Our Office Boy." *Pick-Me-Up*, no. 194 (1892): 189.
48. "Pepper and Salt," *Judy*, July 14, 1886, 20.
49. Thomas Hardy, *Jude the Obscure* (Oxford: Oxford University Press, 2008).
50. Ibid., 53.
51. Ibid.
52. Hardy, *Jude*, 53.
53. Sweet, "Get the Best," 130.
54. Margrit Shildrick, *Leaky Bodies and Boundaries: Feminism, Postmodernism and (Bio)Ethics* (London: Routledge, 1997), 55.
55. Sandra Lee Bartky, "Foucault, Femininity and Patriarchal Power," in *Feminism and Foucault: Reflections on Resistance*, eds. Irene Diamond and Lee Quinby (Boston: Northeastern University Press, 1988), 81.
56. Shildrick, *Leaky Bodies*, 56.
57. J. Capp, "Rewination!" *The Sporting Times*, December 8, 1888, no lines.
58. Jonathan Swift, "A Beautiful Young Nymph Going to Bed," Poetry Foundation, accessed July 2, 2018, https://www.poetryfoundation.org/poems/50580/a-beautiful-young-nymph-going-to-bed.
59. G. W. C., "Good Advice and a Wooden Leg," *Hearth and Home: An Illustrated Weekly Journal for Gentlewomen*, no. 116 (1893): 419.
60. Oliver Wendell Holmes, "The Human Wheel, Its Spokes and Felloes," in *Sounds from the Atlantic*, by Oliver Wendell Holmes (Boston: Ticknor and Fields, 1864), 307.
61. André de Blaumont, "My Fiancé's Glass Eye," in *Short Stories: A Magazine of Select Fiction*, ed. Alfred Ludlow White, trans. Mrs. Huntington Denton, vol. XV (New York: The Current Literature Publishing Co., 1894); Andree de Beaumont, "My Fiancé's Glass Eye," *The Sacred Heart Review* 1, no. 1 (1895): 16.
62. de Beaumont, "My Fiancé's Glass," 16.
63. See Jennifer Esmail, "'The Little Dog Is Only a Stage Property': The Blind Man's Dog in Victorian Culture," *Victorian Review* 40, no. 1 (2014).
64. de Beaumont, "My Fiancé's Glass," 16.
65. David Serlin, "Engineering Masculinity: Veterans and Prosthetics after World War Two," in *Artificial Parts, Practical Lives: Modern Histories of Prosthetics*, eds. Katherine Ott, David Serlin, and Stephen Mihm (New York: New York University Press, 2002), 52.

66. Charlotte Brontë, *Jane Eyre*, 3rd ed. (London: Norton, 2001). Of course, from a feminist or critical disability-studies perspective it is easy to critique the morality of Rochester's actions leading up to the fire—keeping Bertha Mason locked in his attic being his most egregious—but to a Victorian reader his attempt to rescue her was clearly intended be read as heroic and/or redemptive. His impairments serve a double narrative function as embodied punishments for his past transgressions and emblems of his renewed chivalric character.
67. John Reynders & Co., *Illustrated Catalogue and Price List of Surgical Instruments, Spectacles, Eyeglasses, Optical Goods, Orthopaedical Apparatus, Trusses, Supporters,* Etc. Etc. (New York: John Reynders & Co., 1889), 73; Williamson, "Seven Ways."
68. Charles Lenz & Sons, *Charles Lenz & Sons' Illustrated Catalogue and Price List of Surgical Instruments, Hospital Supplies, Orthopaedical Apparatus, Trusses,* Etc., 4th ed. (Philadelphia: Charles Lenz & Sons, 1892), 78; Williamson, "Seven Ways."
69. Charles Truax, Greene & Co., *Price List of Physicians' Supplies* (Chicago: Charles Truax, Greene & Co., 1893), 1174 and 1162; Williamson, "Seven Ways."
70. Down Bros., *A Catalogue of Surgical Instruments and Appliances Manufactured and Sold by Down Bros* (London: Down Bros., 1890), 464; Lawrence H. Officer and Samuel H. Williamson, "Five Ways to Compute the Relative Value of a UK Pound Amount, 1270 to Present," Measuring Worth, accessed July 3, 2018, www.measuringworth.com/uscompare/.
71. Lawrence H. Officer and Samuel H. Williamson, "Annual Wages in the United States, 1774–Present," Measuring Worth, accessed July 6, 2018, http://www.measuringworth.com/USwages/; Gregory Clark, "What Were the British Earnings and Prices Then? (New Series)," Measuring Worth, accessed July 6, 2018. http://www.measuringworth.com/ukearncpi/.
72. Richard Corson, *Fashions in Eyeglasses* (London: Peter Owen, 1980), 114.
73. Marius Hentea, "Monocles on Modernity," *Modernism/Modernity* 20, no. 2 (2013).
74. Peter Stoneley, *Mark Twain and the Feminine Aesthetic* (Cambridge: Cambridge University Press, 1992), 32.
75. Charles Dickens, *Barnaby Rudge* (Oxford: Oxford University Press, 2003); Charles Dickens, *Dombey and Son*, Project Gutenberg, last modified September 25, 2016, https://www.gutenberg.org/files/821/821-h/821-h.htm.
76. Dickens, *Dombey*, ch. 32.
77. Ibid., ch. 4.
78. Ibid., ch. 4, 17, 50.

79. Ibid., ch. 9.
80. Ibid., ch. 15.
81. Ibid., ch. 32.
82. Ibid., ch. 39.
83. Ibid., ch. 23.
84. Ibid.
85. Jerrold, "Eyes," 65.
86. Dickens, *Dombey*, ch. 9.
87. Ibid., ch. 15.
88. Schaffer, *Romance's Rival*, 196.
89. Dickens, *Dombey*, ch. 60.
90. John Jordan, "Openings" (lecture, Dickens Universe 2016, University of California, Santa Cruz, July 31, 2016).
91. Geoffry Brandon, "The Story of the Jewel in the Stage Coach as Related by Simon Simple Barrister-at-Law," *The Ludgate* 8 (1899).
92. Ibid., 540–42.
93. "He Fixed Her with His Glassy Eye," *Cheshire Observer*, April 5, 1879, 3.
94. Clark, "What Were."
95. Vanessa Warne, "'To Invest a Cripple with Peculiar Interest': Artificial Legs and Upper-Class Amputees at Mid-Century," *Victorian Review* 35, no. 2 (2009): 97.
96. Ibid., 94.
97. [Smith], "The Lucky," 380.
98. M. Sullivan, *Stronger Than Death*, ch. 17–20, *The New Monthly Magazine* 146, no. 593 (1870): 492.
99. Heinrich Zschokke, "Das Bein," *Erheiterungen* 1 (1811).
100. [Heinrich Zschokke], "The Leg—a True Story," *The Kaleidoscope: or, Literary and Scientific Mirror* 7, no. 365 (1827).
101. [Heinrich Zschokke], "The Leg," *Bentley's Miscellany* 3 (1838); [Heinrich Zschokke], "The Leg," *Chambers's Edinburgh Journal*, no. 376 (1839); [Heinrich Zschokke], "The Leg," *The Penny Satirist*, no. 375 (1844); [Heinrich Zschokke], "The Leg," *Bristol Mercury*, January 24, 1857; [Heinrich Zschokke], "The Leg," *Santa Fe New Mexican*, April 4, 1899; [Heinrich Zschokke], "Was He a Madman?", *Bow Bells: A Magazine of General Literature and Art for Family Reading* 19, no. 484 (1873); [Heinrich Zschokke], "Losing a Leg to Gain a Wife," *Hampshire Telegraph and Sussex Chronicle* Etc., April 4, 1885; [Heinrich Zschokke], "Love's Amputation; or, A Leg for a Lady," *Hampshire Telegraph and Sussex Chronicle* Etc., March 7, 1891; [Heinrich Zschokke], "Giving a Limb for a Wife, with Future Regrets," *Bossier Banner*, April 16, 1891; [Heinrich Zschokke], "A Case in Surgery. Which Shows That a Man May Do Too Much for Love of Woman," trans. E. C. Waggener, *National Tribune*, February 2, 1893.

102. [Heinrich Zschokke], "The Leg—a True Story," *Wilmingtonian, and Delaware Advertiser*, May 24, 1827, front page.
103. Michael Oliver, *The Politics of Disablement* (Basingstoke: Macmillan Education, 1990).
104. Alison Kafer, *Feminist, Queer, Crip* (Bloomington: Indiana University Press, 2013), 2.
105. [Zschokke], "The Leg—a True Story," front page.
106. [Robert Williams Buchanan], "The Fleshly School of Poetry," *Contemporary Review* 18 (1871): 334–50.
107. Robert Williams Buchanan, "Lady Letitia's Lilliput Hand," *Temple Bar* 4 and 5 (1862), 552.
108. Michie, *The Flesh*, 98.
109. Clare Stainthorp, "Activity and Passivity: Class and Gender in the Case of the Artificial Hand," *Victorian Literature and Culture* 45, no. 1 (2017): 9.
110. David Bolt, "Not Forgetting Happiness: The Tripartite Model of Disability and Its Application in Literary Criticism," *Disability & Society* 30, no. 7 (2015).
111. Buchanan "Lady Letitia's," 131.
112. Henry Heather Bigg, *On Artificial Limbs, Their Construction and Application* (London: John Churchill, 1855).
113. Pamela K. Gilbert, "The Will to Touch: David Copperfield's Hand," *19: Interdisciplinary Studies in the Long Nineteenth Century*, October 2014, http://19.bbk.ac.uk/articles/10.16995/ntn.695/; Stainthorp, "Activity."
114. Martha Stoddard Holmes and Mark Mossman, "Disability in Victorian Sensation Fiction," in *A Companion to Sensation Fiction*, ed. Pamela K. Gilbert (Chichester: Wiley-Blackwell, 2011), 493–94.
115. Bolt, "Not Forgetting."
116. Kafer, *Feminist*, 3.
117. Wilkie Collins, *Hide and Seek* (Oxford: Oxford University Press, 2009); Wilkie Collins, *Poor Miss Finch* (Oxford: Oxford University Press, 2008).
118. Schaffer, *Romance's Rival*, 160–98.

References

Alcott, Louisa May. *Behind a Mask: or, A Woman's Power*. In *Louisa May Alcott Unmasked: Collected Thrillers*, edited by Madeleine Stern, 361–429. Boston: Northeastern University Press, 1995.

A. M. "Was She False?" *The London Reader: Of Literature, Science, Art and General Information* 26, no. 656 (1875): 75–76.

Armstrong, Nancy. *Desire and Domestic Fiction: A Political History of the Novel*. Oxford: Oxford University Press, 1987.

Bartky, Sandra Lee. "Foucault, Femininity and Patriarchal Power." In *Feminism and Foucault: Reflections on Resistance*, edited by Irene Diamond and Lee Quinby, 61–86. Boston: Northeastern University Press, 1988.

Bashford, Alison. *Purity and Pollution: Gender, Embodiment, and Victorian Medicine*. Studies in Gender History. Basingstoke: Palgrave Macmillan, 1998.

Beaumont, Andree de. "My Fiancé's Glass Eye." *The Sacred Heart Review* 1, no. 1 (1895): 16.

Bigg, Henry Heather. *On Artificial Limbs, Their Construction and Application*. London: John Churchill, 1855.

Blaumont, André de. "My Fiancé's Glass Eye." In *Short Stories: A Magazine of Select Fiction*, edited by Alfred Ludlow White, translated by Mrs. Huntington Denton, vol. XV: 277–81. New York: The Current Literature Publishing Co., 1894.

Boissonneau, Auguste. *General Observations on Artificial Eyes, Their Adaption, Employment and the Means of Procuring Them*. Paris: J.-B. Baillière and Son, 1862.

Bolt, David. "Not Forgetting Happiness: The Tripartite Model of Disability and Its Application in Literary Criticism." *Disability & Society* 30, no. 7 (2015): 1103–17.

Boumelha, Penny. "Introduction." In *The Woodlanders*, by Thomas Hardy, xi–xxvi. Oxford: Oxford University Press, 2009.

Brandon, Geoffry. "The Story of the Jewel in the Stage Coach as Related by Simon Simple Barrister-at-Law." *The Ludgate* 8 (1899): 536–44.

Brontë, Charlotte. *Jane Eyre*. 3rd ed. London: Norton, 2001.

Buchanan, Robert Williams. "Lady Letitia's Lilliput Hand." *Temple Bar* 4 and 5 (1862): 551–69; 114–31.

[Buchanan, Robert Williams]. "The Fleshly School of Poetry." *Contemporary Review* 18 (1871): 334–50.

Capp, J. "Rewination!" *The Sporting Times*, December 8, 1888.

Charles Lenz & Sons. *Charles Lenz & Sons' Illustrated Catalogue and Price List of Surgical Instruments, Hospital Supplies, Orthopaedical Apparatus, Trusses, Etc.* 4th ed. Philadelphia: Charles Lenz & Sons, 1892.

Charles Truax, Greene & Co. *Price List of Physicians' Supplies*. Chicago: Charles Truax, Greene & Co., 1893.

"He Fixed Her with His Glassy Eye". *Cheshire Observer*, April 5, 1879.

Clark, Gregory. "What Were the British Earnings and Prices Then? (New Series)." Measuring Worth. Accessed July 6, 2018. http://www.measuringworth.com/ukearncpi/.

Collins, Wilkie. *Hide and Seek*. Oxford: Oxford University Press, 2009.

———. *Poor Miss Finch*. Oxford: Oxford University Press, 2008.

Corson, Richard. *Fashions in Eyeglasses*. London: Peter Owen, 1980.

"Damages for an Eye: What Would Compensate for Mental Anguish and Disability?" *New York Times*, December 4, 1909.

Davis, Tracy C. "Actresses and Prostitutes in Victorian London." *Theatre Research International* 13, no. 3 (1988): 221–34.

Dickens, Charles. *Barnaby Rudge*. Oxford: Oxford University Press, 2003.

———. *Dombey and Son*. Project Gutenberg. Last modified September 25, 2016. https://www.gutenberg.org/files/821/821-h/821-h.htm.

Dickie, Simon. "Hilarity and Pitilessness in the Mid-Eighteenth Century: English Jestbook Humor." *Eighteenth-Century Studies* 37, no. 1 (2003): 1–22.

———. *Cruelty and Laughter: Forgotten Comic Literature and the Unsentimental Eighteenth Century*. Chicago: University of Chicago Press, 2011.

Dodd, William. "A Narrative of the Experience and Sufferings of William Dodd, A Factory Cripple. Written by Himself." In *Factory Lives: Four Nineteenth-Century Working-Class Autobiographies*, edited by James R. Simmons Jr., 181–300. London: Broadview, 2007.

Down Bros. *A Catalogue of Surgical Instruments and Appliances Manufactured and Sold by Down Bros*. London: Down Bros., 1890.

Drake, John S. *Drake's Patent Artificial Legs, Hands, Arms, &c.* Boston: J. Drake, 1859.

Ellis, Havelock. *Man and Woman: A Study of Human Secondary Sexual Characters*. London: Walter Scott, 1904.

Esmail, Jennifer. "'The Little Dog Is Only a Stage Property': The Blind Man's Dog in Victorian Culture." *Victorian Review* 40, no. 1 (2014): 18–23.

"The False Hair". *Chambers's Journal of Popular Literature, Science and Arts* 8, no. 396 (1861): [65]–67.

Foster, Shirley. *Victorian Women's Fiction: Marriage, Freedom, and the Individual*. London: Routledge, 2012.

Gilbert, Pamela K. *Disease, Desire and the Body in Victorian Women's Popular Novels*. Cambridge: Cambridge University Press, 1997.

———. "The Will to Touch: David Copperfield's Hand." *19: Interdisciplinary Studies in the Long Nineteenth Century*, October 2014. http://19.bbk.ac.uk/articles/10.16995/ntn.695/.

Griffin, Ben. *The Politics of Gender in Victorian Britain: Masculinity, Political Culture and the Struggle for Women's Rights*. Cambridge: Cambridge University Press, 2012.

G. W. C. "Good Advice and a Wooden Leg." *Hearth and Home: An Illustrated Weekly Journal for Gentlewomen*, no. 116 (1893): 419.

Haley, Bruce. *The Healthy Body and Victorian Culture*. Cambridge: Harvard University Press, 1978.

Hardy, Thomas. *The Woodlanders*. Oxford: Oxford University Press, 2009.

———. *Jude the Obscure*. Oxford: Oxford University Press, 2008.

Hentea, Marius. "Monocles on Modernity." *Modernism/Modernity* 20, no. 2 (2013): 213–37.

"Her Fatal Sneeze". *Illustrated Chips*, no. 2 (1890): 26.

Holborn Dental Institute. "Advertisement for the Holborn Dental Institute." 1886. British Library Online Gallery. Accessed July 5, 2018. http://www.bl.uk/onlinegallery/onlineex/evancoll/a/014eva000000000u05139000.html.

Holmes, Martha Stoddard. *Fictions of Affliction: Physical Disability in Victorian Culture*. Ann Arbor: University of Michigan Press, 2004.

Holmes, Martha Stoddard, and Mark Mossman. "Disability in Victorian Sensation Fiction." In *A Companion to Sensation Fiction*, edited by Pamela K. Gilbert, 493–506. Chichester: Wiley-Blackwell, 2011.

Holmes, Oliver Wendell. "The Human Wheel, Its Spokes and Felloes." In *Sounds from the Atlantic*, by Oliver Wendell Holmes, 282–327. Boston: Ticknor and Fields, 1864.

Hood, Thomas. *Miss Kilmansegg and Her Precious Leg*. In *Selected Poems of Thomas Hood*, edited by John Clubbe, 193–270. Cambridge: Harvard University Press, 1970.

Jerrold, William Blanchard. "Eyes Made to Order." *Household Words* 4, no. 81 (1851): 64–66.

John Reynders & Co. *Illustrated Catalogue and Price List of Surgical Instruments, Spectacles, Eyeglasses, Optical Goods, Orthopaedical Apparatus, Trusses, Supporters, Etc. Etc.* New York: John Reynders & Co., 1889.

Jordan, John. "Openings." Lecture to Dickens Universe 2016. University of California, Santa Cruz. July 31, 2016.

Kafer, Alison. *Feminist, Queer, Crip*. Bloomington: Indiana University Press, 2013.

"Kitty the Careless". *Judy*, August 1, 1883.

LaCom, Cindy. "'Is It More than Lame': Female Disability, Sexuality, and the Maternal in the Nineteenth-Century Novel." In *The Body and Physical Difference: Discourses of Disability*, edited by Sharon L. Snyder and David T. Mitchell, 189–201. Ann Arbor: University of Michigan Press, 1997.

"The Lame Landlord's Story". *Temple Bar* 20 (1867): 127–34.

"Made-up Beauty". *The Botanico-Medical Recorder* 8, no. 2 (1839): 30.

Mangham, Andrew, and Greta Depledge. *The Female Body in Medicine and Literature*. Liverpool: Liverpool University Press, 2013.

Marcus, Sharon. *Between Women: Friendship, Desire, and Marriage in Victorian England*. Princeton: Princeton University Press, 2007.

Maudsley, Henry. "Galstonian Lecture II on Relations between Body and Mind, and between Mental and Other Disorders of Nervous System." *Lancet*, April 30, 1870, 609–10.

McAleavey, Maia. *The Bigamy Plot: Sensation and Convention in the Victorian Novel*. Cambridge: Cambridge University Press, 2015.

Michie, Elsie B. *The Vulgar Question of Money: Heiresses, Materialism, and the Novel of Manners from Jane Austen to Henry James.* Baltimore: Johns Hopkins University Press, 2011.

Michie, Helena. *The Flesh Made Word: Female Figures and Women's Bodies.* Oxford: Oxford University Press, 1987.

———. *Victorian Honeymoons: Journeys to the Conjugal.* Cambridge: Cambridge University Press, 2006.

———. *Love among the Archives.* Edinburgh University Press, 2015.

Mihm, Stephen. "'A Limb Which Shall Be Presentable in Polite Society': Prosthetic Technologies in the Nineteenth Century." In *Artificial Parts, Practical Lives: Modern Histories of Prosthetics*, edited by Katherine Ott, David Serlin, and Stephen Mihm, 282–99. New York: New York University Press, 2002.

Mr. Foley, Surgeon Dentist. "Advertisement for Mr. Foley, Surgeon Dentist." c.1896. Wellcome Collection. Accessed July 5, 2018. https://wellcomecollection.org/works/j625k4vh.

O'Connor, Erin. *Raw Material: Producing Pathology in Victorian Culture.* Durham: Duke University Press, 2000.

Officer, Lawrence H., and Samuel H. Williamson. "Annual Wages in the United States, 1774–Present." Measuring Worth. Accessed July 6, 2018. http://www.measuringworth.com/USwages/.

Officer, Lawrence H. "Five Ways to Compute the Relative Value of a UK Pound Amount, 1270 to Present." Measuring Worth. Accessed July 3, 2018. www.measuringworth.com/uscompare/.

Oliver, Michael. *The Politics of Disablement.* Basingstoke: Macmillan Education, 1990.

Ott, Katherine. "Hard Wear and Soft Tissue: Craft and Commerce in Artificial Eyes." In *Artificial Parts, Practical Lives: Modern Histories of Prosthetics*, edited by Katherine Ott, David Serlin, and Stephen Mihm, 147–70. New York: New York University Press, 2002.

"Pepper and Salt", *Judy*, July 14, 1886.

Perry, Ruth. *Novel Relations: The Transformation of Kinship in English Literature and Culture, 1748–1818.* Cambridge: Cambridge University Press, 2004.

Poe, Edgar Allan. "The Spectacles." In *Prose Tales (Second Series) Arthur Gordon Pym*, by Edgar Allan Poe, 161–98. Boston: Dana Estes, 1884.

Poovey, Mary. *Uneven Developments the Ideological Work of Gender in Mid-Victorian England.* Chicago: University of Chicago Press, 1988.

Schaffer, Talia. *Romance's Rival: Familiar Marriage in Victorian Fiction.* Oxford: Oxford University Press, 2016.

Serlin, David. "Engineering Masculinity: Veterans and Prosthetics after World War Two." In *Artificial Parts, Practical Lives: Modern Histories of Prosthetics*, edited by Katherine Ott, David Serlin, and Stephen Mihm, 45–74. New York: New York University Press, 2002.

Shanley, Mary Lyndon. *Feminism, Marriage, and the Law in Victorian England.* Princeton: Princeton University Press, 1993.

Shildrick, Margrit. *Leaky Bodies and Boundaries: Feminism, Postmodernism and (Bio)Ethics.* London: Routledge, 1997.

[Smith, Sarah]. "The Lucky Leg." *Household Words* 19, no. 469 (1859): 374–80.

Stainthorp, Clare. "Activity and Passivity: Class and Gender in the Case of the Artificial Hand." *Victorian Literature and Culture* 45, no. 1 (2017): 1–16.

Stone, Lawrence. *The Family, Sex and Marriage in England 1500–1800.* London: Penguin Books, 1990.

Stoneley, Peter. *Mark Twain and the Feminine Aesthetic.* Cambridge: Cambridge University Press, 1992.

Storey, John. *The Making of English Popular Culture.* London: Routledge, 2016.

Stratz, Carl Heinrich. *Die Schönheit des weiblichen Körpers.* 14th ed. Stuttgart: F. Enke, 1903.

"The Stricken Fawn". *Fun* 58, no. 1489 (1893): 219.

Sullivan, M. *Stronger Than Death.* Ch. 17–20. *The New Monthly Magazine* 146, no. 593 (1870): 489–512.

Sweet, Ryan. "'Get the Best Article in the Market': Prostheses for Women in Nineteenth-Century Literature and Commerce." In *Rethinking Modern Histories of Prostheses in Anglo-American Commodity Cultures, 1850–1960,* edited by Claire L. Jones, 114–36. Manchester: Manchester University Press, 2017.

Swift, Jonathan. "A Beautiful Young Nymph Going to Bed." Poetry Foundation. Accessed July 2, 2018. https://www.poetryfoundation.org/poems/50580/a-beautiful-young-nymph-going-to-bed.

"Too Hard upon My Aunt". *All the Year Round* 10, no. 242 (1863): 381–84.

Vrettos, Athena. *Somatic Fictions: Imagining Illness in Victorian Culture.* Stanford: Stanford University Press, 1995.

Walker, Alexander. *Intermarriage; or the Mode in Which, and the Causes Why, Beauty, Health, and Intellect Result from Certain Unions, and Deformity, Disease and Insanity from Others.* London: John Churchill, 1838.

Warne, Vanessa. "'To Invest a Cripple with Peculiar Interest': Artificial Legs and Upper-Class Amputees at Mid-Century." *Victorian Review* 35, no. 2 (2009): 83–100.

Watt, Ian. *The Rise of the Novel: Studies in Defoe, Richardson, and Fielding.* Berkeley: University of California Press, 1974.

Williamson, Samuel H. "Seven Ways to Compute the Relative Value of a U.S. Dollar Amount, 1774 to Present." Measuring Worth. Accessed July 19, 2021. www.measuringworth.com/uscompare/.

"Women. By Our Office Boy". *Pick-Me-Up*, no. 194 (1892): 189.

Wood, Ellen. *East Lynne.* Oxford: Oxford University Press, 2008.

Zschokke, Heinrich. "Das Bein." *Erheiterungen* 1 (1811): 87–100.

[Zschokke, Heinrich]. "The Leg—a True Story." *The Kaleidoscope: or, Literary and Scientific Mirror* 7, no. 365 (1827a): 427–28.

———. "The Leg—a True Story." *Wilmingtonian, and Delaware Advertiser*, May 24, 1827b.

———. "The Leg." *Bentley's Miscellany* 3 (1838): 480–84.

———. "The Leg." *Chambers's Edinburgh Journal*, no. 376 (1839): 94–95.

———. "The Leg." *The Penny Satirist*, no. 375 (1844): 4.

———. "The Leg." *Bristol Mercury*, January 24, 1857.

———. "Was He a Madman?" *Bow Bells: A Magazine of General Literature and Art for Family Reading* 19, no. 484 (1873): 367.

———. "Losing a Leg to Gain a Wife." *Hampshire Telegraph and Sussex Chronicle Etc.*, April 4, 1885.

———. "Giving a Limb for a Wife, with Future Regrets." *Bossier Banner*, April 16, 1891a.

———. "Love's Amputation; or, A Leg for a Lady." *Hampshire Telegraph and Sussex Chronicle Etc.*, March 7, 1891b.

———. "A Case in Surgery. Which Shows That a Man May Do Too Much for Love of Woman." Translated by E. C. Waggener. *National Tribune*, February 2, 1893.

———. "The Leg." *Santa Fe New Mexican*, April 4, 1899.

CHAPTER 6

Signs of Decline? Prostheses and the Ageing Subject

Hail thou! who liest so snug in this old box;
With sacred awe I bend before thy shrine!
Oh! 't is not clos'd with glue, nor nails, nor locks,
And hence the bliss of viewing thee is mine.
Like my poor aunt, thou haft seen better days!
Well curl'd and powder'd. once it was thy lot
Balls to frequent, and masquerades, and plays,
And panoramas, and the Lord knows what!
Oh! thou hast heard e'en Madam Mara sing,
And oft-times visited my Lord Mayor's *treat;*
And once, at court, wast notic'd by the King,
Thy form was so commodious, and so neat.
Alas! what art thou now? a mere old mop!
With which our housemaid Nan, who hates a broom,
Dusts all the chamber in my little shop,
Then slyly hides thee in this lumber-room!
Such is the fate of wigs! *and* mortals *too!*
After a few more years than thine are past,
The Turk, the Christian, Pagan, and the Jew,
Must all be shut up in a box *at last!*
Vain man! *to talk so loud, and look so big!*
How small's the difference 'twixt thee and a wig!
How small indeed! for speak the truth I must,
Wigs *turn to* dusters, *and* man *turns to* dust.

R. Sweet, *Prosthetic Body Parts in Nineteenth-Century Literature and Culture*, Palgrave Studies in Nineteenth-Century Writing and Culture, https://doi.org/10.1007/978-3-030-78589-5_6

—"Modern Sonnet. To an Old Wig," in *The Spirit of the Public Journals for 1802*, vol. 6 (London: James Ridgway, 1803).

This poem, which appeared numerous times in the nineteenth-century periodical press, provides an apt starting point for a chapter about prosthesis use and ageing.[1] Published first in 1802, these verses dedicated "To an Old Wig" put forward several themes that would be developed and drawn upon repeatedly in depictions of elderly prosthesis users over the course of the century. Such topics include the elision of age and prosthesis use, the revulsion to vanity and falsehood, and the tension between the natural and the artificial. The poem equates the aged with redundancy, an association that would be buttressed by many writings in the years that followed. Although we see the prostheses of the elderly depicted as somewhat useless devices in many other depictions, here an old wig has more use (and longevity) than an elderly person. While man "turns to *dust*," wigs "turn to *dusters*." Old wigs, these stanzas suggest, retain at least a degree of purpose and functionality, while elderly people are treated as detritus.

Although, as Pat Thane observes, in the nineteenth century "people aged over sixty were a smaller proportion of the population than for several centuries," a significant share of the prosthesis users that we see depicted in nineteenth-century fictional and cultural works are elderly.[2] There are several famous examples: Brigadier General John A. B. C. Smith from Edgar Allan Poe's "The Man That Was Used Up" (1839), Captain Cuttle and Mrs. Skewton in Charles Dickens's *Dombey and Son* (1846–1848), and Mr. Bashwood and Mrs. Milroy in Wilkie Collins's *Armadale* (1864–1866).[3] From these select works alone, we see a range of users, including elderly spinsters, ageing aesthetes, and maimed war veterans. These individuals use a variety of devices, from wigs, false teeth, and makeup to artificial legs, hook hands, and even mechanical palates. Despite their differences, all of these characters are bound by plots that centre on decline, deception, and discrepancy (between old and young, and between artificial and organic).

While attempts to resist senescence today are often applauded, during the nineteenth century, the use of prostheses by those experiencing physical decline in order to hide signs of ageing was regularly lambasted as an act of vanity or, worse, fraudulence. Though these attitudes were commonplace in the literature of this period, ageing prosthesis users and artificial body parts themselves performed a variety of roles in fictional works, functioning interchangeably as comic motifs, as signifiers for a variety of character traits—including vanity, duplicitousness, and credulousness—and as synecdoches for a broader social system that privileged youth and physical wholeness. Many imaginaries satirized this social privileging, showing how it led to preposterous results. As I show here, the association of ageing and prosthesis use was so strong that even those who might not be considered elderly in terms of years lived were depicted as aged or past their best because of their use of a prosthetic device. As we might expect, primarily cosmetic prostheses, including wigs, false teeth, and artificial eyes, became synonymous with representations of ageing.

Attitudes to Ageing

To consider first some facts and figures, George R. Boyer and Timothy P. Schmidle's study "Poverty among the Elderly in Late Victorian England" shows us that in 1861 "there were 932,000 persons aged 65 and over in England and Wales, representing 4.6 per cent of the population. By 1891, there were nearly 1.4 million persons aged 65 and over in England and Wales, or 4.7 per cent of the population."[4] As Anne-Julia Zwierlein, Katharina Boehm, and Anna Farkas suggest in the introduction to their important collection *Interdisciplinary Perspectives on Aging in Nineteenth-Century Culture*, "the percentage of older people among the population was less pronounced during the nineteenth century." However, as they show, numerous factors contributed to what they describe as "an unprecedented level of cultural attention on the experience of aging and old age."[5] Zwierlein, Boehm, and Farkas point to factors such as the expanding market for print publications, social reform movements that laid the groundwork for the emergence of the welfare state, and the growth of new scientific and medical modes of inquiry (including gerontology). Karen Chase argues similarly that cultural attention to the elderly reached a peak at the end of the nineteenth century.[6]

The elderly certainly came to cultural prominence in the Victorian period but it is harder to pinpoint exactly what the prevailing attitudes to old age were. As Thane explains: "In all times there have been competing optimistic and pessimistic paradigms of old age and we cannot readily determine whether one or the other was culturally dominant. All cultures have a variety of images of ageing available to them from which individuals and groups shape their expectations. These images shift and compete and if any one of them gains hegemony it does not necessarily do so for long."[7]

As Chase shows us, attitudes to ageing were split by several tensions. First, the continuing resonance of the Ciceronian ideal of old age as a period of dignity, serenity, wisdom, respect, and self-respect contrasted growing examples of old age as misery, bewilderment, loneliness, and disentitlement. Second, the spreading deprivation of the aged at the bottom of the social ladder opposed the power of ageing figures including Queen Victoria, William Ewart Gladstone, and Benjamin Disraeli. Third, cross-generational relationships were a preoccupation of imaginative life while generational fractures also became apparent, especially when partisans of the New contested the iconography of the old—in particular at the moment of the Queen's Jubilees and in the retrospection of the *fin de siècle*.[8] While this chapter resists a monolithic view of ageing in the nineteenth century, many literary responses to the prostheses of the elderly cast a negative light on both the aged and the means of artifice that they used to mask their perceived bodily losses. Such responses, however, ironically often attacked the very devices brought to life by privileging youth and wholeness.

The nineteenth-century prosthesis market itself drew together concepts of ageing and physical loss in its advertising rhetoric. Cashing in on the insecurities of those showing signs of physical decline, contemporary prosthesis makers marketed their devices towards those wishing to disguise losses resulting from senescence. An advertisement for Mr. Scott, Surgeon and Mechanical Dentist, published in *Bell's Life in London and Sporting Chronicle* in 1830, for instance, began, "IMPORTANT to all desirous of Perfection, Comfort and Economy."[9] In the advertisement section of a single 1858 copy of *Lady's Magazine*, a periodical written for well-to-do mature women, two adverts for false teeth appear next to adverts for hair dye, hair serum, and a baldness cure.[10] An 1867 *Bell's Life in London and Sporting Chronicle* advertisement for Messrs. Mosley's "Teeth and Painless Dentistry" assured potential customers, many of whom we assume were middle aged or older, that "artificial teeth (from one to a complete set) can be inserted with so exact a semblance to nature that detection is impossible, the original proportions of the face and mouth being *restored* to their pristine perfection" (emphasis added).[11] Significant here is the promise to *restore* "pristine perfection," which aligns youth with wholeness and therefore perfection. Intriguingly, cultural depictions of users of false teeth and other prostheses tended not to agree with the bold promises of advertisements. The obviously false and thus abhorrent prosthesis became a popular shorthand for the ageing subject across a range of literary and cultural sources.

Curiously, it was not only devices that we might call primarily cosmetic prostheses, such as wigs and dentures, which became insignias for old age in this period; peg legs, rather than their more expensive and sophisticated counterparts, artificial legs (see Chap. 4), were also often associated with old age. This association stemmed in part from the proliferation of images of prosthesis-using Greenwich and Chelsea pensioners (see Figs. 6.1 and 6.2), and other maimed war veterans, who were sometimes represented as decrepit street beggars (see Figs. 4.1 and 4.2).[12] Due to the kind of amputation procedures performed on many maimed military men and the relatively basic provisions provided by the state, as well as the perpetuation of attitudes that saw disability as a sign of unfitness for work, British veterans

Fig. 6.1 A lithograph depicting four Greenwich pensioners congregating around a table. The pensioner on the left is an arm amputee and the figure to the right uses a pin-design prosthetic leg. *Greenwich*, n.d., lithograph, 14 × 19.7 cm, Wellcome Trust, London. Image courtesy of the Wellcome Collection. https://wellcomecollection.org/works/cr2enpd4. CC BY 4.0

Fig. 6.2 A late nineteenth-century etching of Chelsea pensioners playing draughts. The figure in the foreground is seated, with his wooden leg protruding to the side of the table. George Fox, "A pair of Pensioners," seated at a table, playing draughts, with others looking on, inside the Royal Hospital, Chelsea, with a vignette of tricorn hat, wooden leg, clay pipe, drinking glass and medals at lower left, 1887, etching, 30.5 × 23.8 cm, Wellcome Collection, London. Image courtesy of the Wellcome Collection. https://wellcomecollection.org/works/sacuezvv. CC BY 4.0

tended to be jobless after sustaining their injuries, reducing them to a state similar to that of elderly men, who were also often excluded from the workplace.[13] Legislation such as the Poor Law Amendment Act 1834 reinforced associations of the elderly and the disabled, since this legislation exempted from the workhouse unemployed subjects from both groups. In terms of masculinity, then, infirmity caused by impairment and infirmity caused by old age were conceptually linked in as much that both states were believed to represent an inability to work. For women, by comparison, the use of false limbs had a similar ageing effect since attitudes towards conspicuous physical incompleteness excluded them from the marriage market, aligning them conceptually with so-called spinsters.

Given that the perceived absurdity of prosthesis use is a focus of my analysis, it is perhaps unsurprising that humour is a prevalent aspect of the sources under discussion. This chapter understands humour primarily by way of incongruity theory, the current pre-eminent approach in humour studies. As John Morreall explains, "What makes ... situation[s] ... humorous, according to the Incongruity Theory, is that there is something odd, abnormal or out of place, which we enjoy in some way. In its simplest form, the theory says that humorous amusement is the enjoyment of incongruity."[14] Foundational humour studies writer Stephen Leacock sees Victorian humour as emblematic of this theory: "[it] finds its basis in the incongruity of life itself, the contrast between the fretting cares and the petty sorrows of the day and the long mystery of tomorrow."[15] My discussion bears in mind the compatibility of the incongruity theory of humour with Mitchell and Snyder's theory "narrative prosthesis," a comity exposed by Tom Coogan and Rebecca Mallet.[16] Underlining the overlap in these approaches, Coogan and Mallet write: "[n]arrative prosthesis identifies disability as the crutch upon which narratives lean for their representational power; incongruity theory, as Morreall explains, attributes humour to situations where 'there is something odd, abnormal or out of place, which we enjoy in some way.'"[17] What both theories have in common, then, is a notion that the strategic placement of the non-normative is a key component of affect.

In the sources that I examine here, the prostheticized body or prosthetic part is the incongruity at the comic centre, a perceived oddness that is emphasized by the age of the subject. The use of a prosthetic device is sometimes revealed suddenly in an unveiling scene that uses a period of suspense beforehand to heighten comic release—providing further evidence for the compatibility of incongruity and relief models of humour.[18]

In these instances, the unveiling scene often occurs at a moment or in a specific setting that exacerbates the incongruousness of the prosthesis and therefore the embarrassment of the user. Elsewhere, the prosthetic body part is centred relatively early and it is the incongruousness of it, as signalled by the extent to which it stands out—because it looks unreal, does not suit the appearance of the user, or is being used in a non-normative way—that draws readers' laughs. Another comic trope draws humour from the incongruousness of a non-prosthesis user's inability to distinguish the artificial from the real, a distinction that is apparently blatant to everyone else, including the reader. In disability-studies terms, one might argue that each representational type draws attention to "a pattern of cultural representation which always maintains physically different people as other, as alien, as the object of curiosity or hostility or pity, rather than as part of the group."[19] However, behind the ableist and ageist veneer of comical representations of ageing prosthesis users was often a critique of the privileging of the young and whole body, physical states that appear both ephemeral and unstable. These depictions draw our attention to the ludicrous situations and displays that cultural pressures to maintain normative standards give rise to.

The following analysis traces the trope of the ageing prosthesis user from the 1830s through to the fin de siècle. I begin by investigating Poe's representations of grotesque artificially constructed elderly subjects, portrayals that drew from the harsh humour of eighteenth- and early nineteenth-century caricature and satire, redirecting laughs towards those unable to perceive the difference between the artificial and the organic. I then go on to show how ageing prosthesis users were reconceptualized in sensation fiction, a genre that borrowed much from the Gothic mode that underpinned Poe's fiction. In Wilkie Collins's narratives, wigs are used simultaneously as light comic props, which at surface value further stigmatize the elderly and physically incomplete, and synecdoches through which wider social structures are critiqued. Meanwhile, in the late nineteenth-century periodical press the trope of the ageing prosthesis user had become a popular culture icon, one that was less vulgar than earlier deployments but still critical of the absurdities that social pressures to conform to hegemonic norms could bring about. Finally, I explore several late-century imperial adventure fictions, a genre that utilized the prevalent prosthetic motif. Revealing an increased consciousness of age anxiety, imperial adventure tales continued to laugh at ageing prosthesis consumers by transgressively turning preferences for wholeness on their head.

Vanity and Calamity

Images of the elderly undressing and removing their prostheses have a long history. There were popular humorous tropes during the "golden age of caricature," which extended from the 1760s through to the 1820s, a period in which artists such as James Gillray, Thomas Rowlandson, and George Cruikshank achieved notable success.[20] In P. Roberts's 1807 coloured etching after G. M. Woodward, *Celia Retiring*, a bald, toothless, and overall dishevelled-looking elderly woman gives precise instructions to her servant about laying out her wig, false bosom, dentures, and false eyebrows so that she can get ready to meet "[her] Lord Ban" in good time the next day (Fig. 6.3). Similarly, F. S. Delpech's 1825 coloured lithograph after L. L. Boilly, *Les époux assortis*, displays an elderly couple removing their various prostheses—wigs, sets of false teeth, and a glass eye (Fig. 6.4). An even earlier but remarkably similar literary example of such a scene is described in detail in Jonathan Swift's infamous satirical poem "A Beautiful Young Nymph Going to Bed" (1731).[21] Dickie shows us that this period was remarkably spiteful towards those seen as physically aberrant. The body, in its most unsightly, vulnerable, and non-normative forms, was, as Dickie argues, a central theme in eighteenth-century humour despite the period's demure pretensions.[22] In Boilly, Woodward, and Swift, the ageing body is exposed as frail and abhorrent without prostheses. The prostheses themselves are simultaneously uncanny, deceitful, and concealing devices, which align with a grotesque aesthetic. The elderly are unceremoniously mocked while the culture of artifice is lambasted.

Though by no means a Victorian invention, the trope of the elderly person with prostheses extended and perpetuated from the 1830s onwards. Chase shows how the theme of ageing became increasingly central in the social consciousness as the nineteenth century progressed. As Chase suggests, a constellation of events brought about "the ripening consciousness of age," which she calls "the invention of the elderly subject."[23] These factors include the legislative movement towards the Old Age Pensions Act 1908; the ageing and then the death of the Queen Victoria; the "ethological" investigations conducted by Charles Booth; the emergence of gerontology as a medical sub-discipline; the increasing perception of a generational divide; and the proliferating images of ageing bodies. The increase in cultural representations of the elderly with prosthetics maps roughly onto Chase's trajectory, which traces the ways in which the elderly become a mainstream concern at the fin de siècle. We can add four

Fig. 6.3 A coloured etching depicting an elderly lady providing strict instructions to her servant regarding her prostheses. She says, “Molly, mind what I say to you. Lay my wig on the top of the drawers. Take care of my bosom and don’t rumple it. Lay my eye in the dressing box, and the row of teeth by the side of it. And call in again for my eyebrows. Lay everything in such a manner that I may easily find them in the morning, as I wish to be made up by 12 precisely, in order to meet my Lord Bandash.” P. Roberts after G. M. Woodward, *Celia Retiring*, c. 1807, etching with watercolour, 35 × 26.5 cm, Wellcome Collection, London. Image courtesy of the Wellcome Collection. https://wellcomecollection.org/works/m4kdj55c. CC BY 4.0

Fig. 6.4 A coloured lithograph portraying an elderly couple removing their prostheses—wigs, eyes, and teeth. F. S. Delpech after Louis Leopold Boilly, *Les epoux assortis*, 1825, lithograph with watercolour, Wellcome Collection, London. Image courtesy of the Wellcome Collection. https://wellcomecollection.org/works/f8p4baxv. CC BY 4.0

additional factors to Chase's comprehensive list. First, Britain witnessed the entering into old age of its amputee veterans of the Napoleonic wars and a continued cultural presence of war veterans, including Greenwich and Chelsea pensioners. Second, the 1850s American invention of sulphur-hardened rubber, Vulcanite, which became the most popular material for manufacturing the bases of dentures, inaugurated "the era of false teeth for the masses."[24] Third, wig-wearing took off in the 1860s and 1870s as women sought to emphasize "their femininity through masquerade."[25] Finally, in ironic contrast to the previous two contexts, post-Regency puritanism extended into the nineteenth century, labelling the use of artifice a vanity.[26]

Poe provides a couple of examples of grotesque elderly characters, which evoke those depicted in the earlier works of Woodward, Boilly, and Swift. Both "The Man That Was Used Up" and "The Spectacles" (1844) depict aged characters whose bodies are substantially supplemented by prostheses.[27] Like Swift's "Young Nymph," the artificiality of the prosthesis-using characters described in Poe's stories is revealed in undressing scenes of sorts. In "The Man that Was Used up," Brevet Brigadier General John A. B. C. Smith, a retired and venerated war veteran, appears mysterious to the tale's curious narrator until the latter witnesses him getting dressed one morning. The veteran is as much an assemblage of many components as his excessively long name: several of his body parts are prosthetic, including a leg, an arm, both shoulders, his bosom, his hair, his teeth, an eye, and his palate. Despite his extreme dependence on prostheses, Smith is earlier described as striking in appearance: "There was an air *distingué* pervading the whole man, which spoke of high breeding, and hinted at high birth."[28] Likewise, in "The Spectacles" Madame Eugénie Lalande is presented as an enticingly beautiful physical specimen. Moreover, like Smith, she turns out to be the user of many prostheses. Eugénie's "perfect fullness and *tournure*" turns out to be the work of a "false *tournure*"; her "beautiful black hair," the work of a wig; and her "pearly teeth," the work of dentures.[29] These devices allow Eugénie to trick the narrator, Simpson, who it is revealed is the attractive woman's great-great-grandson. Eugénie agrees to marry Simpson to teach him a lesson about the importance of vision as a physiognomic tool of selection. Her rather heavy-handed moral is that Simpson should wear spectacles to make up for his poor eyesight. As in "The Man that Was Used up," the ability of prostheses to mimic a younger, natural physical appearance provides the elderly user with a specific kind of technologically enabled agency.

One could argue that "The Spectacles" is a story all about the importance of keeping up-to-date with technologies that enhance or supplement the body, since the titular assistive aid ultimately makes Simpson better off—at the very least less prone to making erroneous misreadings about those around him.

These imaginaries, which can both be read as reworkings of the mistaking-artifice-for-reality narrative "The Sandman" by E. T. A. Hoffmann (1817), reveal anxieties about the increasing dominance of technology mixed with distrust towards deceitful ageing subjects who rely on prostheses to attain a youthful aesthetic.[30] These stories poke fun at society's reliance on bodily technologies while problematizing the social pressure to use such devices. The practice of prostheticizing the body is displayed as grotesque by the dismantling and rebuilding scenes of both works while those unable to recognize artifice are mocked for their ignorance. As much as Madame Lalande's and Smith's respective prostheses are shown to be curiously incongruous when they are detached, physiognomic adjustment is shown to be needed in a world where appearances can be misleading. Relating to nineteenth-century physiognomy and visual modes of judging today, Sherrona Pearl has shown us that "[w]hat one sees in others tells a great deal about oneself."[31] The humour in these stories stems in part from what we learn about the narrator through his physiognomic judgements: that he is dangerously (but also hilariously) unperceptive. Critically, however, in both stories the privileging of youth and wholeness is shown to be self-defeating since the pressure that it produces leads the elderly and physically incomplete to prostheticize their bodies, thereby duping the normate into believing that incomplete bodies are actually physically whole. In these specific cases, the ability of the prosthesis to enhance the appearance of the subject complicates the hegemony of normalcy, since the false takes precedence over the organic: prostheses achieve a better-than-real aesthetic for Poe's cyborgian elderly subjects, thus confounding the subject group who are supposed to be dominant. The normate becomes a victim of his own prejudices.

False Part, Flawed Whole

Drawing from the Gothic, sensation fiction is another genre that offers much to the discussion of bodies, ageing, and artifice in the Victorian period. The human body is central to the narrative structure and affective force of sensation fiction. As Martha Stoddard Holmes and Mark Mossman write: "Sensation fiction's relationship to embodiment has been overdetermined from the start, given that its poetics, its plotting and characterization, and its critical reception have used the body as a nexus of expression, experience, and meaning-making."[32] The portrayals that I discuss here comically mock vain, ageing users of artifice while subtly delivering a double assault on both social pressures to conform to normate physical standards and the linked matrimonial preferences for good looks and wealth that such a society propagated.

In his novels *Armadale* (1864–1866), *The Law and the Lady* (1875), and *The Black Robe* (1881), the sensation novelist Wilkie Collins depicts a variety of ageing male and female characters who the author mocks for their vanity.[33] Lisa Niles makes a convincing argument about the way that cosmetics are deployed in Collins's *Armadale* as a false concern that draws the reader's attention to what she identifies as the real threat, embodied by Lydia Gwilt: a criminal, ageing body that successfully passes as a younger body in a marriage market that relies on age being clearly demarcated. According to Niles, "Collins challenges the terms upon which society constructs a middle-class, marriageable female identity."[34] Collins's critique of Victorian's society's paradoxical engagement with ageing, however, is not restricted to women. As Collins suggests through his depictions of unmarried ageing male characters—such as Bashwood from *Armadale* and Major Fitz David from *The Law and the Lady*—single ageing (usually middle-class) men could also feel the brunt of social prejudices against incomplete bodies, inclining them to use artificial adornments. Nonetheless, Collins's attitude to prosthesis-using ageing men is far from sympathetic. His fictions critique both the culture that encourages ageing men (and women) to use cosmetic aids (including artificial body parts) to look younger and fuller, and the men whom he sees as foolish enough to give into the weight of societal pressures. Both *Armadale* and *The Law and the Lady* underline what Collins saw as the futility of male vanity by suggesting that physical appearances are not as important for men as they might think; in these narratives, wealth is critically exposed as the fundamental, though problematic, criterion for women when choosing their partners. Collins thus displays consternation towards two related social systems: first, that which pressures ageing men to use cosmetic adornments, thereby in his eyes making themselves look ridiculous; and second, the marriage system in which appearances and wealth are privileged over deeper personal qualities.

Collins's novels depict conventional responses to artifice as a means of disguising senescence. His two female characters who conspicuously use prostheses are Mrs. Milroy from *Armadale* and Miss Notman from *The Black Robe*. Mrs. Milroy, the paranoiac and invalid mother of Miss Neelie Milroy (the eventual wife of the protagonist Allan Armadale), is discussed at some length by Niles. She considers by the wig- and rouge-wearing Mrs. Milroy a character who reaffirms prejudices against elderly women attempting to defy the process of ageing.[35] To put this assertion into context, Lola Montez, Countess von Landsfeld, advised in 1858 that "in no case can even rouge be used by ladies who have passed the age of life when roses are natural to the cheek. A rouged old woman is a horrible sight—a distortion of nature's harmony!"[36] Niles, however, contends that Mrs. Milroy "provides readers with a false sense of surety in the ability to detect artifice, as her self-presentation appears to reduce the threat of cosmetics from something fearful to something ridiculous."[37]

While Mrs. Milroy is both married and impaired by illness, which makes her vanity (in Victorian terms) even more preposterous, Miss Notman from *The Black Robe* is single but also presented as absurd for using a wig—in this case, especially, it would appear, because of her lowly status as a housekeeper. The narrator's first significant description of Notman reveals that her use of a wig is a translucent insignia of her flawed character:

> When Miss Notman assumed the post of housekeeper in Lady Loring's service, she was accurately described as "a competent and respectable person;" and was praised, with perfect truth, for her incorruptible devotion to the interests of her employers. On its weaker side, her character was represented by the wearing of a youthful wig, and the erroneous conviction that she still possessed a fine figure. The ruling idea in her narrow little mind was the idea of her own dignity. Any offence offered in this direction oppressed her memory for days together, and found its way outwards in speech to any human being whose attention she could secure.[38]

The syndetic pairing of wearing "a youthful wig" and "the erroneous conviction that she still possessed a fine figure" puts her very much in the same boat as vain, ageing prosthesis-using women depicted elsewhere in Victorian fiction, including Lady Maria Esmond from Thackeray's *Virginians* (1857–1859), Mrs. Skewton from Dickens's *Dombey and Son*, and Lady Carbury from Anthony Trollope's *The Way We Live Now* (1875).[39] However, unlike Lady Maria, Mrs. Skewton, Lady Carbury, or even Mrs. Milroy, Miss Notman is of a lower social station, making her vanity seem redundant. Whereas upper-class and middling female characters, including Lady Maria, Lady Carbury, and Mrs. Skewton, face increased public scrutiny in their respective imaginative universes as respectable women of polite society, thereby pressuring them to attempt to look more youthful, Miss Notman's work confines her to domestic spaces, thereby necessitating a neat and tidy but not showy appearance. To a contemporary readership, her supposed vanity thus exposes her as someone with ideas above her station. This trait combined with the view that she is elderly makes her an object of ridicule.[40]

Miss Notman's "youthful wig" to her matches what she believes is a "fine" youthful figure. This stance positions her among the many who tried (and often failed) to disguise visible signs of ageing, believing that looking younger would bring rewards. Unfortunately for Miss Notman, she is not Lydia Gwilt—a character able to disguise her true age. Notman's attempt to look youthful is mocked by Collins, who takes a stance not too dissimilar to Margaret Morganroth Gullette, whose work problematizes the logic of "performing" youth through cosmetic surgery in the twenty-first century:

> The men and women who get facelifts or liposuction … try on a crude appliance to pass for "younger" to stereotyped eyes. Whatever else is wrong with this, it's like stuffing in falsies or a pair of socks to represent gender. It's both bad acting and a misreading of culture. They are treating "age"—or in this case "youth"—as a wholly separable identity. Obliterating a few of the obvious signals that everyone knows comes solely from the age code, like "wrinkles" or "love handles," they try to forget that in our culture the hypercritical age gaze notices all the more obsessively their other decline-linked signals.[41]

For Collins, Notman's "youthful wig" is also a result of "bad acting" and "a misreading of culture." Her wig is conspicuous: through a contrast to her otherwise ageing appearance, it draws attention to her as an imposter. She misreads her culture not only by thinking that people will be duped by her ploy but also by presuming that it is necessary for her to maintain an appearance of youth in the first place. However, when one considers the overwhelming social pressure that encouraged women to display youthful and physically whole appearances, we begin both to empathize with Miss Notman and to see Collins's deeper critique of social pressures regarding physical presentation.

The most poignant aspect of Collins's depiction of Miss Notman is the way that her youthful wig is used as a symbol of her weaknesses. Notman is revealed to be both frivolous and pompous by her actions: she unwittingly reveals information to the novel's villain, Father Benwell, that allows him to conspire against Collins's protagonist. She also almost quits her position after feeling that her dignity is affronted following a disagreement regarding whether to serve an oyster omelette before or after pudding. The early conviction that "her character was represented by the wearing of a youthful wig, and the erroneous conviction that she still possessed a fine figure" thus stands true.[42] Collins's past-their-best female prosthesis users seem to provide historical evidence for Susan Sontag's "double standard" of ageing.[43] These depictions suggest that in the high-Victorian period, some ageing women felt the need to look younger, even when in social positions where it might seem unnecessary.

Collins's critique of prosthesis users for attempting to pass as youthful, however, extended beyond women. Through his depiction of foolish, ageing men, Collins associated the problem that he saw not with a specific gender but with the social system more broadly. Major Fitz David from *The Law and the Lady* is another comic character, though his eventual fate exposes the sorry results of an ageing aesthete's use of cosmetics. When the novel's protagonist and narrator, Valeria Woodville, an attractive but intelligent and driven young woman, first approaches the Major for help uncovering the mystery behind her husband's decision to conceal his true surname (Macallan), the Major is described in flamboyant terms:

> [A] well-preserved old gentleman of, say, sixty years old, little and lean, and chiefly remarkable by the extraordinary length of his nose. After this feature, I noticed next his beautiful brown wig; his sparkling little gray eyes; his rosy complexion; his short military whisker, dyed to match his wig; his white teeth and his winning smile; his smart blue frock-coat, with a camellia in the button-hole; and his splendid ring, a ruby, flashing on his little finger as he courteously signed to me to take a chair.[44]

A little later, we learn that the Major's eyebrows are also "dyed to match his whiskers."[45] Fitz David's impressively made-up, pristine appearance is concomitant with his youthful, showy persona. At one point, the Major is described as "speaking in the character of a youth of five-and-twenty."[46]

In addition to his faux-young appearance, he also has a taste for women somewhat younger than he is. One of his admirers, the girl whom he eventually marries, is described by Valeria as "a plump, round-eyed overdressed girl, with a florid complexion and straw colored hair."[47] Valeria herself is described as "young" by the Major's servant, Oliver, priming the former for his first encounter with Collins's heroine:

> "Is she young, Oliver?"
> "Yes, sir."
> "And—pretty?"
> "Better than pretty, sir, to my thinking."
> "Aye? aye? What you call a fine woman—eh, Oliver?"
> "Certainly, sir."
> "Tall?"
> "Nearly as tall as I am, Major."
> "Aye? aye? aye? A good figure?"
> "As slim as a sapling, sir, and as upright as a dart."
> "On second thoughts, I am at home, Oliver. Show her in! show her in!"[48]

Here and elsewhere, the Major's weakness for women, a characteristic usually associated with younger men, appears ridiculous.

Despite the fact that artificial hair sales soared in the years leading up to the serialized publishing of *The Law and the Lady*, Major Fitz David belongs to an era when the male wearing of wigs was often treated with comic disdain.[49] As John Woodforde observes in *The Strange Story of False Hair* (1971), for men in the nineteenth century, "[w]igs, like false teeth, were considered a shameful vanity."[50] In *Punch*, an article appeared in 1855 that mocked a new West End hairdresser who had recently launched a "New Wig Club," which opened its door to only the "titled, the wealthy, and persons of fashion."[51] *Punch* exposed this "fashionable" clientele as elderly, vain, and entirely pompous through a slanderous list of imagined rules for the club:

> 1. The New Wig Club is instituted for the purpose of promoting the privacy of persons who wear wigs or ornamental hair, or who resort to the dyeing process to conceal their greyness.
>
> 2. The New Wig Club shall consist of any number of members, who shall be either grey or bald, and any one with black hair who is not bald will be black-balled.
>
> 3. Ladies and gentlemen whose hair is beginning to fall off or turn grey may be admitted as honorary members for one month, after which they must either purchase a wig or a bottle of hair-dye, in order to continue to enjoy the privilege of admission.

> 4. Each candidate for admission shall be proposed by one member who is bald or grey, and seconded by another; and a lock of the candidate's hair, or if bald, a curl of his wig, shall be hung up for at least one week before the day of the election in the Club-room.
>
> 5. No member shall be allowed to vote at an election whose hair has not been dyed, or his wig dressed, within one month from the day of voting.
>
> 6. No wash or dye except that supplied by the Club, shall be made up in the Club on any pretence whatever.
>
> 7. No member shall bring a stranger into the Club on any pretence whatever.
>
> 8. The Club shall be open for the dyeing and hair-dressing of members from ten in the morning until ten at night, except during the London season, when the Club shall be open till midnight.
>
> 9. Any defect in a wig or a hair-dye, must be complained of to the Manager of the Club; and if a head is badly dressed, or not done to the turn of a hair, the complaining member may put the curl on the Secretary's box, which must be kept under lock till the complaint is verified.[52]

The emphasis in this piece is very much on the grey and the bald rather than the stylish and/or fashionable, thereby equating wig use with ageing and a condition of supposed desperation. Rule number seven draws our attention to the commercial aspect of wig use, exacerbating the stranglehold that prosthetists had over those unhappy with their incomplete physical appearances.

A much earlier poem, which appeared in the aptly named magazine the *Age* in 1832, expressed more direct concerns about the way that greedy wig sellers exploited ageing men. "The Old Man and His Grey Wig" tells the story of "An old gentleman, bless'd with a good crop of hair," whose barber "insisted a wig he must wear."[53] The barber disingenuously tells the "old" man that he will lose his hair without the aid of false hair. When the elderly man dismisses the barber's advice, the latter ironically claims that using a wig will save the gentleman hairdressing expenses. This promise proves, unsurprisingly, to be an utter lie. The "old" man eventually agrees to shave his head and purchase a grey wig, but after a year he is shocked when an extortionate bill arrives, including costs for

> bleeding, for purge, cauterizing,
> For pills anti-*bil*-ions, and black drafts to swig,
> Head-shaving, and trimming, and wig-modernizing,
> And lengthy pig-tail to the fusty *grey wig*.[54]

The poem exposes what the anonymous speaker sees as a ruthless exploitation of those susceptible to being conned: in this case, elderly men.

While Collins's depiction of Major Fitz David is arguably less sympathetic to the victim of such a morally bankrupt system, the author does appear to be troubled by a related and, to him, equally immoral financial concern: the concept of marrying for money. In the final chapter of *The Law and the Lady*, we learn that the Major has successfully married, but we soon realize that his success had little to do with his faux-youthful looks. His new wife, the "plump, round-eyed overdressed girl" aforementioned, reveals to Valeria that she married the Major for financial security:

> [I]t was a great deal easier to get the money by marrying the old gentleman. Here I am, provided for—and there's all my family provided for, too—and nothing to do but to spend the money. I am fond of my family; I'm a good daughter and sister—*I* am! See how I'm dressed; look at the furniture: I haven't played my cards badly, have I? It's a great advantage to marry an old man—you can twist him round your little finger. Happy? Oh, yes![55]

Valeria, no doubt like many Victorian readers of this novel, is disgusted by this young woman's conduct. She states: "When a woman sells herself to a man, that vile bargain is none the less infamous (to my mind), because it happens to be made under the sanction of the Church and the Law."[56] However, the larger point that Collins is getting to here goes beyond highlighting the attitude of this particular character as morally reprehensible. For Collins, society more widely is at fault. This woman's behaviour is undesirable but painfully savvy: she has herself and her family to look after. For her, the easiest and (ironically) most respectful way to achieve financial security is to marry a wealthy, and thus often older, man. Still, the married Major is a shadow of his former self, both physically—looking "hopelessly and undisguisedly" an "old" man—and behaviourally—following his marriage. For example, Valeria describes how the Major "looked at [his wife] submissively between every two words that he addressed to [her]."[57] The Major's financial status makes him a target for women whose equivalents today are labelled by some as "gold diggers," while his position as an elderly bachelor makes him apparently desperate for the attention of women. This combination, Collins shows, lays the foundations for what the reader imagines will be an unhappy marriage.

Although Collins's scornful tone makes it clear that it is not the Major's physical attractiveness that makes him, in the words of Miserrimus Dexter, an "elderly human lap-dog" for women, the Major's wig is a significant aspect of his appearance.[58] After all, the Major's wig is not just any old wig: it is "a beautiful brown wig," suggesting that it is of a standout quality. Writing two decades earlier, Alexander Rowland wrote that "a peruke, containing only three ounces of hair … is frequently sold at the price of twenty-five to thirty shillings."[59] Considering that average annual nominal earnings in 1851 were £33.58, meaning that a wig according to Rowland cost on average over double average weekly earnings, it becomes clear that the Major's wig is a sign of affluence—even if one considers the higher average annual nominal earnings of £53.86 when *The Law and the Lady* was published in 1875, twenty-five to thirty shillings remained a relatively high commodity price.[60]

Despite its high commodity value, in *Armadale* Bashwood's youthful wig does not signify nearly enough wealth for Collins's ambitious femme fatale anti-heroine Lydia Gwilt. Here, Collins's infamous anti-heroine sets her sights on marrying Allan Armadale, inheritor of the large estate Thorpe Ambrose. When her plan to marry Armadale fails, she concocts another plan to gain possession of the property, this time by way of marriage to Ozias Midwinter, whose real name is also Allan Armadale (he is, in fact, among four Allan Armadales mentioned in the novel). While putting her designs into action, Lydia Gwilt attracts the help of the elder former steward of Thorpe Ambrose, Mr. Bashwood, who is infatuated with her. Bashwood performs duties as a spy for Gwilt. Over the course of his infatuation with Gwilt, Bashwood's wig and false teeth, signifiers of his old age and illegitimacy as a viable match for Gwilt, become indexes of the intensity of his passion. These devices are also comic motifs through which Collins again critiques the social pressures that drive individuals to present themselves in this way.

When we first meet Bashwood, he is described in the following manner:

> He was a lean, elderly, miserably respectable man. He wore a poor old black dress-coat, and a cheap brown wig, which made no pretence of being his own natural hair. Short black trousers clung like attached old servants round his wizen legs; and rusty black gaiters hid all they could of his knobbed, ungainly feet. Black crape added its mite to the decayed and dingy wretchedness of his old beaver hat; black mohair in the obsolete form of a stock drearily encircled his neck and rose as high as his haggard jaws. The one morsel of colour he carried about him was a lawyer's bag of blue serge, as lean and limp as himself. The one attractive feature in his clean-shaven,

> weary old face was a neat set of teeth—teeth (as honest as his wig) which said plainly to all inquiring eyes, "We pass our nights on his looking-glass, and our days in his mouth."[61]

As with Miss Notman's wig, we see Bashwood's false hair and teeth reflect his personality: "honest," "cheap," and unconvincing. We sense that even though he uses devices that would be considered vestiges of vanity, he has more likely than not been tricked or otherwise coerced into using these primarily cosmetic prostheses: his appearance is otherwise a little on the shabby side of neat and rather reserved, suggesting that he is not vain per se.

Amid his humble attire, Bashwood's anomalous prostheses provide a subtle reference, on the one hand, to the social pressure to look young and physically complete, and, on the other hand, to the persuasiveness of contemporary commercial prosthesis rhetoric. The conspicuousness of Bashwood's prostheses, considered alongside his weak persona, imply that he may have fallen victim to the misleading marketing of prostheses. But when we read Collins's next description of Bashwood, it becomes even clearer that his prostheses are analogous to his feebleness as a character:

> There, perched comfortless on the edge of his chair, sat the poor broken-down, nervous wretch, in his worn black garments, with his watery eyes, his honest old outspoken wig, his miserable mohair stock, and his false teeth that were incapable of deceiving anybody—there he sat, politely ill at ease; now shrinking in the glare of the lamp, now wincing under the shock of Allan's sturdy voice; a man with the wrinkles of sixty years in his face, and the manners of a child in the presence of strangers; an object of pity surely, if ever there was a pitiable object yet![62]

Here, Collins emphasizes Bashwood's unfortunateness through a series of oxymoronic descriptions. The elderly steward has an "outspoken wig" and "false teeth that were incapable of deceiving anyone." Though he is "a man with the wrinkles of sixty years," he has the "manners of a child in the presence of strangers."[63] His appearance thus fails him on several accounts: his prostheses are blatant and his haggard face—often a symbol of experience, self-assuredness, and sagacity—contrasts his timid, withdrawn, and childlike behaviour. Bashwood's prostheses do not enable him to pass as youthful and whole; instead, they expose him as an elderly man trying to look vigorous.

The appearance of Bashwood's prostheses, however, changes significantly after he has run errands for his beloved Miss Gwilt. By this point, Bashwood's affections have matured to a stage whereby he idolizes the ground that Gwilt walks on. For instance, when catching sight of her prior to a rendezvous, Bashwood mutters to himself: "I wish I was the ground she treads on! I wish I was the glove she's got on her hand!" His prostheses, as well as his appearance in general, are presented in more refined but nonetheless ridiculous terms: "His personal appearance had been apparently made the object of some special attention. His false teeth were brilliantly white; his wig was carefully brushed; his mourning garments, renewed throughout, gleamed with the hideous and slimy gloss of cheap black cloth. He moved with a nervous jauntiness, and looked about him with a vacant smile."[64] Here, in an ironic shift, Bashwood's finely groomed prostheses accentuate the artificiality and grotesqueness of his appearance. His dentures appear "brilliant white," a colour of teeth that few would have achieved in this period.[65] While many strove for white teeth, as we can see from the burgeoning trade of the teeth-whitening-powder in this period, few managed to maintain "pearly whites"—due, in part, to the often damaging role played by falsely heralded whitening powders, which tended to be so gritty that they wore away tooth enamel as well as stains.[66] Those who did achieve teeth whiteness from a rigorous hygiene routine, or the use of artifice, were not always depicted in the most positive lights. Mr. Carker from Dickens's *Dombey and Son*, for example, is described at regular intervals as bestial because of the immaculate whiteness of his teeth. Dickens's depiction can be considered in relation to contemporaneous depictions of white-toothed vampire figures, such as Sir Frances Varney from James Malcom Rymer's popular penny dreadful *Varney the Vampire* ([1845–1847]).[67] Bashwood's white teeth thus not only appear out of place in context with his ageing face but also signify an almost parasitic element to his character, associating him with earlier vampire and vampire-like figures, including Varney and also Heathcliff in Emily Brontë's *Wuthering Heights* (1847).[68] The notion of a creepy, elderly, white-toothed man who preys on younger women is thus conjured, but ultimately it is Bashwood who is the victim of Gwilt's almost mesmeric influence over him.

Above all, Bashwood's prostheses provide the reader with an index for the elderly steward's desperate passion for Gwilt. As his prostheses become more polished, and ironically more visible, his feelings towards Gwilt also intensify. When Bashwood eventually realizes that he is being played, he "lift[s] his youthful wig a little from his bald old head," emphasizing its symbolic closeness to his passions.[69] To use Bill Brown's framework, here "we begin to confront the thingness" of Bashwood's wig: the object "stops working" for Bashwood, thereby further emphasizing its "thingness" to the reader.[70] The "thingness" of the device was brought to our attention much earlier, since it never functioned effectively, but at this point the narrative places a strong focus on it. The wig's pretence as a natural part of Bashwood is obliterated through action. The lifting of the wig is an act of resignation. Bashwood's attempts to look younger are no longer needed as his chances with Gwilt, though non-existent to begin with, become clear to him at last. The torment that he experiences is simultaneously both cruelly humorous and pitiful, underlining Collins's complex satire of the social pressures that might encourage ageing men to act like Bashwood.

Losing One's Wig

Collins was hardly the only Victorian writer to make fun of ageing prosthesis users. The late nineteenth-century periodical press used the prosthesis mishap as a major trope. This trope, though popular from early in the century, really took off from the 1870s onwards, reflecting developments in journalistic style and prosthesis manufacture. The "New Journalism" that emerged in the 1870s and 1880s, which brought an investigative and sensational tone to the exposure of scandals and intrigues, responded to the increasingly competitive marketplace by becoming more attuned to consumers' preferences for stories about popular culture. As Mark Hampton notes, savvy editors recognized that readers "cared less about Home Rule or Bulgarian atrocities" than about "football scores, divorce cases, murder trials, and fashionable dresses."[71] "As a result," Nicholson writes, "parliamentary reports, financial bulletins and local news items gave way to serialized fiction, household tips, children's pages, gardening advice, poetry, competitions and comic clippings."[72] Alongside these developments, prostheses, such as dentures and wigs, became increasingly prevalent in society. By 1878, for instance, one could purchase a complete set of teeth in vulcanite form from £4 4 s—half the price of a set that had gold uppers.[73]

In the stories that appeared in periodicals from this period, we often see ageing or elderly prosthesis users, whose wigs blow off in the wind, whose false teeth fall out or otherwise malfunction, or whose cover is blown by an innocent observer—often a child or animal. These depictions, like Collins's, at surface level display an unsympathetic attitude to those in physical decline, reinforcing conceptual tensions and divides between the organic and the artificial, and youth and old age. Even though they are presented in a predominantly ableist and often ageist fashion, such stories implicitly challenge the predominant regime of privileging physical wholeness and youth, since they draw our attention to the incongruity of the false body part on the aged subject. Central to these representations from the periodical press is the incompatibility of the human with the technological appendage. My contention is that these representations, while clearly humorous, were not simply frivolous depictions used to stimulate easy laughs. Instead, these engagements use comedy as a vehicle through which to call into question both the proliferation of certain prosthetic technologies and the very logic behind these devices.

The image of a prosthesis falling off an elderly figure or malfunctioning became increasingly popular in the nineteenth century. We see this image replicated numerous times in sources such as Grace Goldney's 1870 serialized novella *Marion's Choice*, in which an elderly man's hat and wig are blown down the street by a strong gust of wind; William Henry Archibald Chasemore's 1878 *Judy* cartoon "Wicklebury's Wig," which depicts a similar situation facing an elderly bachelor who purposely bought a new wig to impress a particular woman; the 1891 *Funny Folks* short story "Mr. Sagtooth's First Appearance," which represents an elderly, wig-wearing man, whose wig falls off during an impromptu theatrical performance, causing a comic stir; and Harold Copping's circa 1895 illustration for F. Scarlett Potter's 1895 novel *Hazelbrake Hollow*, "It caught his wig and jerked it into the air," which portrays another elderly man, whose wig is taken off by the hook of a fishing line.[74] Though decidedly less accidental, the *Fun* cartoon "Retributive Justice" (1879) shows the results of an elderly female false teeth user being hit in the mouth with a snowball: it "disarranged her artificial teeth, so that she couldn't close her mouth, and, thinking she had lockjaw, she sat down in the snow and screamed horribly" (see Fig. 6.5). In each of these cases, the prosthesis is primarily a motif of light amusement. Laughter is stimulated by the irony of the scenarios depicted: a device used solely to disguise physical loss and thereby hide one's "shame" is suddenly removed or displaced, unwittingly exacerbating the embarrassment of the user, who is not only unveiled as incomplete but also exposed as vain and/or fraudulent. The shock, panic, and/or embarrassment of the prosthesis user losing or having damaged his or

Fig. 6.5 A cartoon showing how dentures can malfunction—in this case, because of a direct hit from a snowball. "Retributive Justice," *Fun* 29 (1879): 93. Image published with permission of ProQuest. Further reproduction is prohibited without permission

her artificial body part adds to the hilarity of the scenario. For instance, the facial expression and body language of Wicklebury in Chasemore's cartoon is typically stunned and thus conducive of laughter.

In relation to the human-technology interface, the wig was often depicted as a device that almost never manages to successfully convince those around it that it is natural. As C. D. Shanly wrote in an article for *The Treasure of Literature and the Ladies' Treasury* in 1872: "Few things adopted with intention to deceive are less deceptive than wigs."[75] Its shortcomings in this respect are exacerbated when it falls off. Instances of prosthesis failure suggest that there is too great a discrepancy between the real and the artificial for the latter to successfully mimic and integrate fully with the former. The losing of one's wig shows, on the one hand, a practical dilemma about fixing a wig to an often-bald head, and, on the other hand, a conceptual contrast between the natural and the manmade. Woodforde shows us that those men ordering wigs to disguise baldness were asked to provide very precise measurements to manufacturers in order to ensure a good fit, a natural-looking appearance, the lessening of any peruke slippage.[76] There was also a variety of fixatives available to secure wigs to heads, including springs, strings, pastes, gums, and even the

"Golden Cement":" a product advertisements stated would "stick [a wig] tight on the head, and not move, as if it had grown there."[77] Still, cultural depictions suggest that wigs fell off at an embarrassing rate.

In addition, we also see artificial hair's unconvincingness in scenes where it is noticed and/or removed by seemingly innocent parties, including animals and children. A humorous example appears in G. Renaud's 1887 cartoon "Waggles's Wig."[78] In this story, a vain and ageing dandy visits a woman named Lady Trabazon, clearly with the intention of wooing her and/or her more youthful female companion. Upon his arrival, Lady Trabazon's pet marmoset takes a clear interest in the well-presented man. During Waggles's conversation with Trabazon, the marmoset mischievously steals his wig. Chaos ensues. In "The Story of Gaffer Grey" (1871), an elderly man's wig goes missing after a jackdaw takes it to make a nest with. Meanwhile, in C. E. Pearce's "Princess Prettypet" (1873), a mischievous Princess plays a trick on her music Professor by tying his wig to a sleeping dog, which when awoken by cries of "cats," wakes up and tears the false hair to shreds. The same is the case in "The General's Wig" (1875), where an elderly General's wig is stolen and destroyed by a cockatoo.[79] Along similar lines, in the 1880 *Boys of England* serial *The Good-Natured Boy; or, What Came of It*, a mischievous parrot directs an innocent boy into a matron's room while she is getting changed.[80] As the boy discovers, much to the matron's chagrin, she wears false teeth and a glass eye. She whips him with a wet towel as a punishment.

What these representations suggest is that even supposedly unenlightened animals were able to perceive the anomalousness of a prosthesis on an elderly person. Moreover, these cultural works recall an observation of Mark Twain, who held critical views about humanity when compared to so-called *Higher Animals*. In his 1896 essay "Man's Place in the Animal World," Twain compared humankind's frailty and reliance on technology to the adaptive prowess of "Higher Animals":

> For style, look at the Bengal tiger—that ideal of grace, beauty, physical perfection, majesty. And then look at Man—that poor thing. He is the Animal of the Wig, the Trepanned Skull, the Ear Trumpet, the Glass Eye, the Pasteboard Nose, the Porcelain Teeth, the Silver Windpipe, the Wooden Leg—a creature that is mended and patched all over, from top to bottom. If he can't get renewals of his bric-a-brac in the next world, what will he look like?[81]

Twain's words critique the ways in which humans deal with the physical decline that accompanies old age. The examples that he provides, including ear trumpets, false teeth, and artificial legs, in addition to his choice of verbs to describe the process of fighting senescence—"mended" and "patched"— draw our attention to the view that the human body cannot successfully integrate with technology; at best, its blemishes can be patched up with non-human parts, but even then its reconstitution is but a temporary—and unconvincing—fix. Twain elsewhere mocks the fashion for prostheticizing the body in physical decline. In his semi-autobiographical 1872 travel book, *Roughing It*, Twain describes a group of elderly women who happily lend each other prostheses for social occasions: "[Miss Jefferson] had a glass eye and used to lend it to old Miss Wagner, that hadn't any, to receive company in; it warn't big enough, and when Miss Wagner warn't noticing, it would get twisted around in the socket, and look up, maybe, or out to one side, and every which way, while t' other one was looking as straight ahead as a spy-glass."[82] For Twain the grotesque use, or rather misuse, of artifice exemplified the shabbiness of the human condition.

Twain's depiction of prosthesis misuse brings us to the several tales that focus on the inability of the elderly to use prostheses correctly. In typical, comic fashion, in Alfred Edersheim's 1873 novel *Shorn to the Wind*, a pastor's wife, who has one arm, ironically carries her prosthesis in a shopping basket instead of using it for practical purpose.[83] Similarly, a popular anecdote, one reprinted several times, tells the story of an elderly woman who had been taken to court by an ocularist after she refused to pay for a glass eye.[84] The woman explains that her artificial eye "is not half the use of my wig and artificial teeth, for I cannot see out of it a bit."[85] In another humorous sketch, an elderly woman asks a friend that wears a glass eye why she did not ask for her eye to be made from more transparent glass so that she could see out of it.[86] These short fictions once again open a conceptual divide between the natural and the artificial and, in fact, complicate the cultural association of the ageing with the prosthetic. These examples suggested that elderly people were out of touch with modern innovation and thus unable to use prosthetic devices to their full potential.

Dazzling Devices and Unlikely Heroes

The non-normative use of a prosthetic device by a subject experiencing physical decline is a motif taken up with surprising frequency in late-century imperial adventure stories. As I will show in this final section, these depictions drew from the contemporary cultural mocking of vain elderly prosthesis users, while further complicating the social pressures that encouraged this behaviour by presenting past-their-best prosthesis-using male imperialists as unlikely heroes. The combination of a device that is perceived to be remarkable with an unlikely ageing hero provides a fascinating comic contrast. In several imperial adventure stories, the artificial body parts of ageing imperialists are perceived by non-Western subjects as objects of awe in comical and narratively significant scenes.[87] In Haggard's *King Solomon's Mines* (1885), Captain John Good's false teeth save the lives of him and his comrades after the dentures are seen by some hostile Kukuanaland warriors as signifiers of divinity. Similarly, and perhaps derivatively, McKean's "A Wig and a Wooden Leg" (1886) tells the story of a wig-wearer and a wooden-leg user, who while working for the same imperial trader are saved in separate attacks from aggressive "Indians" thanks to their respective prostheses being perceived as magical. In the *Pick-Me-Up* magazine narrative poem "A Cure for Cannibalism" (1889), an artificial eye saves one Mr. Brown from being eaten by cannibals in Central Africa:

> The savages were thunderstruck, and marvelled at the fellow.
> Who made them turn a colour that approached a greeny-yellow.
> With many gestures indicating the consensus of the meeting,
> That such a clever party was by far too good for eating.[88]

These prostheses have been understood as examples of ethnocentrism—revelatory of a Western arrogance that mocks non-Westerners for having supposedly absurd beliefs regarding, or unenlightened approaches to, technology.[89] Yet, key comic aspects of these stories have been neglected. The humour stems from a couple of factors relating to the intersections of ageing and prosthesis use: namely, from the way that prostheses become symbols of ageing, thus making their users appear unlikely heroes, and from the unlikelihood of the prosthesis itself as a device that incites fear.

Kay Heath's scholarship on midlife provides a critical framework to examine the transitional life stage between youth and old age that the imperial prosthesis users embody. Midlife, though first defined in an English language dictionary in 1895, was very much a part of the social consciousness for much of the nineteenth century[90]: "Whereas old age is depicted as a final stage increasingly associated with the end of certain activities and identities—lessened marriageability, waning sexuality, and retirement from work—midlife ... stress[es] the possibility that [a] character's fortunes may go either way, from acquiescent decline to a sustained youthfulness."[91] Each of the imperialists described tread this liminal stage of ageing as they perform adventurous deeds, displaying youthful vitality, while their bodies show visible signs of decline. Most significantly, it is the "past-one's-best" persona of each character that is key to both the laughter produced by his interactions with prosthesis and the challenging of physical preferences that each story contributes to.

The impression that the four imperialists are "past-one's-best" emerges when we learn that all of them are seasoned individuals. Good was in the navy for seventeen years before being "turned out"—a sign in itself that he is past his prime when the reader meets him.[92] In McKean's story, it is apparent that the wig-wearer is an imperialist of considerable experience as it is reported that he was scalped by Native Americans "several years before" his second encounter with the violent indigenous people.[93] In "A Cure for Cannibalism," we learn that Mr. Brown is a veteran explorer, implying that he is not youthful. Furthermore, after escaping death, the cannibals call him "father."[94]

Each character also displays the physical cost of years of work overseas. Haggard's narrator, Quatermain, describes Good as an extremely neat but "curious man to look at."[95] Good is physically unimpressive in comparison to his companion Sir Henry Curtis, "the biggest-chested and longest-armed man [that Quatermain] ever saw."[96] Good is of a medium height and stout. He is later described as having a "sickly" face by a bewildered Kukuanaland warrior.[97] More pressingly, Good is extremely reliant on technological adornments—not just a pair of false teeth but also an eye glass. Quatermain alludes to Good's heavy reliance on his eye glass when he jokes: "I thought he used to sleep in it, but afterwards I found that this was a mistake."[98] Prostheses are thus indicators of Good's deteriorating physical capacities and thereby his premature ageing.

Equally, these devices are symbols of his midlife status since they are representative of his attempts to stave off the functional and aesthetic signs of ageing, highlighting his liminal status between youth and old age. In this same way, we can see Mr. Brown's use of an artificial eye and the traders' use of prostheses in McKean's sketch as indicators of physical decline—though eye and limb loss are not signifiers of ageing per se, here they imply that the adventurers are "worn out" or "used up" (to use Poe's phrase) by their excursions.[99] In a time when a healthy body was believed a measure of youth, vigour, and self-assertion, perceived prosthesis use became a cultural shorthand for waning vitality, since nineteenth-century concepts of health relied upon an ideal of physical integrity.[100] Thus, the characters described are not old, but they display evident signs of decline. Because their prostheses become visible, the devices become insignias of ageing. To some extent, the characters also show that sometimes it is not the number of years lived but rather how difficult those years have been that determines one's age in cultural terms. These imperial adventurers show signs of wear and tear, undoubtedly because of their physically taxing and mentally stressful lives, which, one could argue, make them appear old beyond their years.

It is therefore the irony of each figure's bizarre success that makes their stories such sources of intrigue and comic amusement. Heath has shown us that as the nineteenth century progressed, male midlife became an increasingly stressful time as growing emphasis was placed on physicality as a fundamental aspect of masculinity.[101] The emergence and popularity of muscular Christianity with its emphasis on physical brawn, which as Norman Vance argues came about amidst a climate of "emerging individual possibility" between the 1832 and 1867 reform bill, is an important context that Heath shows cultivated anxiety about declining physical prowess among men entering midlife.[102] Furthermore, the fact that "manliness became defined by profession and capacity to work" meant that "waning strength and a sense of unfitness for new times feminized older men, excluding them from power."[103] Regarding the empire, John Tosh notes that strength and endurance were seen as essential traits for maintaining colonial rule—a point that clearly relates to each midlife male prosthesis user described in this section.[104] Because each imperialist is a prosthesis user and thus physically incomplete, he appears diminished when considered alongside late Victorian idols of male physicality such as Eugen Sandow or Sir Henry Curtis from Haggard's novel. Sandow and Sir Henry are both imposing and conspicuously physically complete. What is thus amusing, and somewhat transgressive, in the stories under discussion is that unassuming midlife men are depicted as action heroes—in some cases ironically thanks to the very prostheses that signify their physical decline.

Good, once again, provides the best example to illustrate this point. Despite being a poor shot and somewhat of a bumbling fool in general in terms of his interactions with technological devices, in several instances, to the surprise of both his Western comrades and the reader, his actions save the day. For instance, most memorably, his bizarre, half-dressed attire—with half of his face shaven and his white legs exposed—and idiosyncratic use of false teeth ("dragging the top set down and allowing them to fly back to his jaw with a snap") frighten the Kukuanaland people into thinking that he and his friends are spirits who should be worshipped and respected rather than attacked.[105] He also later provides medical attention to himself and his friends after their violent escapades in the Kukuanaland civil war.

These heroic deeds provide an ironic contrast to the numerous points where he is shown to be the weak link out of Haggard's three white adventurers. In Chapter 4, Good "[falls] a victim to his passion for civilised dress," almost getting trampled to death by a wounded elephant after being "cumbered" by his inappropriate choice of trousers. He then falls over after his polished boots give way under him. In the next chapter, while leading his companions through the desert, Good literally stumbles upon a herd of sleeping quagga (a subspecies of zebra), where he is whisked off into the distance on the back of one. In Chapter 14, he is injured in the battle against Twala's army, which causes him to fall into a life-threatening high fever. In Chapter 18, after Gagool entraps the white men in Solomon's treasure chamber, he is the first to give up hope, reasoning, "What is the good in eating? … the sooner we die and get it over the better." We can associate nearly all of Good's shortcomings with stereotypes of ageing. He attempts to look younger (exemplified by his "passions for civilised dress"); he exhibits signs of mental decline (shown by his mishap with the quagga); and he demonstrates inherent pessimism (displayed by his giving up on hope in the treasure chamber scene).[106] It is thus surprising, amusing, and somewhat transgressive in terms of the cultural preoccupation with youth, physical integrity, and strength that a character with these apparent flaws manages to perform the important heroic deeds listed previously.

The fact that it is a pair of false teeth that elicits awe from the Kukuanaland warriors is plainly ironic and therefore comical.[107] In the case of *King Solomon's Mines*, underpinned by an ethnocentric logic, the narrative encourages readers to laugh at the Kukuanaland people because they are unable to perceive the artifice of Good's teeth; they think that the teeth are real and that Good can remove them at will because he possesses magical powers. A final comic aspect of Good, which is linked to his use of false teeth, is his extreme vanity—a trait commonly mocked in relation to ageing in Victorian literature and culture. Many of the comic moments in the story, including the scene in which Good's bizarre, half-dressed appearance convinces the Kukuanaland people that he and his friends are spirits, can be linked to this aspect of his character. His near misses with death, such as when he "[falls] a victim to his passion for civilised dress" can also be linked to his vanity. Good is thus depicted as absurd, as someone who uses cosmetic aids in inappropriate environments—according to Quatermain, he is "the neatest man [he] ever had to do with in the wilderness."[108] But for many Victorians, Good's inclination towards cosmetics would have seemed ridiculous not just because of his decision to use them "in the wilderness"; the ageing and vain were also represented and discussed as a hilariously hopeless group deserving of mockery.

Haggard's portrayal of Good, however, is not simply a cruel slight against those deemed past their best. The character's prosthesis use fits within a wider cultural critique of vanity, and this criticism serves as a subtle, playful, and humorous curveball against the hegemony of youth and ability. Good's characterization therefore complicates the idolatry displayed towards the chiselled and complete bodies of Sir Henry and Umboppa in Haggard's novel. It is well documented that Haggard's adventure story, like its inspiration Robert Louis Stevenson's *Treasure Island* (1881–1882), was written for "boys of all ages."[109] Good's representation as an unlikely hero whose signs of physical decline are all too apparent provided humour but also fantasy appeal and perhaps even hope to his Victorian midlife male readers.

The sources that I have examined in this chapter consolidated a trope that continues to saturate cultural images of the elderly today. Associations of certain forms of prostheses such as dentures with old age remain so prominent and negative that modern advertisements for products, including denture adhesives, almost never feature direct images of false teeth or elderly users.[110] The conspicuous absence of the prosthetic device shows how influential the legacy of humorous and grotesque depictions of these devices and their elderly users has been.

Nineteenth-century depictions satirized not only the ageing users of such devices but a society that invested too heavily in appearances of youth and physical wholeness. There is thus much that we can learn from these portrayals. We need to look at, and perhaps even laugh at, our own social prejudices to ensure that the lives of the elderly who choose to use prostheses remain free from stigma.

Notes

1. This poem appeared in *New England Quarterly, Spirit of the Public Journal*, and *Sporting Magazine* in 1802. It then appeared in full, abridged, or adulterated form in the following: *Monthly Review* in 1803, *Gentleman's Magazine* in 1804, *Lady's Monthly Magazine* in 1807, *Satirist* in 1837, and *Penny Satirist* in 1838.
2. Pat Thane, "Epilogue," in *Interdisciplinary Perspectives on Aging in Nineteenth-Century Culture*, eds. Anne-Julia Zwierlein, Katharina Boehm, and Anna Farkas (New York: Routledge, 2014), 232–33.
3. Edgar Allan Poe, "The Man That Was Used Up: A Tale of the Late Bugaboo and Kickapoo Campaign," in *The Works of the Late Edgar Allan Poe*, by Edgar Allan Poe, vol. 4 (New York: Blakeman & Mason, 1859); Charles Dickens, *Dombey and Son* (London: Penguin, 2002); Wilkie Collins, *Armadale* (Oxford: Oxford University Press, 1999).
4. George R. Boyer and Timothy P. Schmidle, "Poverty among the Elderly in Late Victorian England," *The Economic History Review* 62, no. 2 (2009): 250.
5. Anne-Julia Zwierlein, Katharina Boehm, and Anna Farkas, eds., *Interdisciplinary Perspectives on Aging in Nineteenth-Century Culture* (New York: Routledge, 2014), 1.
6. Karen Chase, *The Victorians and Old Age* (Oxford: Oxford University Press, 2009).
7. Pat Thane, *Old Age in English History: Past Experiences, Present Issues* (Oxford: Oxford University Press, 2002).
8. Chase, *The Victorians*, 6.
9. Mr. Scott, Surgeon and Mechanical Dentist, Advertisement, *Bell's Life in London and Sporting Chronicle* 417 (1830): 2.
10. Multiple classified advertisements, *Lady's Newspaper*, no. 599 (1858): 399.
11. Messrs. Mosley's Teeth and Painless Dentistry, Advertisement, *Bell's Life in London and Sporting Chronicle* 2, no. 365 (1867): 3.
12. For representations of aged wooden-leg-using Greenwich and Chelsea pensioners, see, for example, Gruff and Glum in Charles Dickens, *Our Mutual Friend* (London: Everyman, 2000); "Laid up at Greenwich,"

Judy: or The London Serio-Comic Journal, December 2, 1868; "The Rewards of Warfare," *Kind Words: A Week-Day Magazine for Boys & Girls*, November 24, 1870; "A Strange Complaint," *Fun*, February 18, 1871. For elderly peg-leg-using beggars, see "Thrift, or Nothing Is Useless," *Chambers's Edinburgh Journal* 1, no. 38 (1844); [Richard H. Horne], "Dust; or Ugliness Redeemed," *Household Words* 1, no. 16 (1850); and the article "Mr. Wegg and His Class," *All the Year Round* 18, no. 441 (1877).

13. Regarding military surgery, Henry Robert Heather Bigg wrote, "many a soldier comes home with what is surgically a splendid stump, but a most inappropriate one mechanically, and to my knowledge it has frequently been found advisable by medical officers attached to the military hospitals at home to re-operate … in order to get better ones." Henry Robert Heather Bigg, *Artificial Limbs, and the Amputations Which Afford the Most Appropriate Stumps in Civil and Military Surgery* (London: Henry Robert Heather Bigg, 1885), 103. In terms of monetary provisions, half a century earlier Henry S. Richardson observed that Privates housed at Greenwich Hospital received 1 s pocket money per week in addition to "lodging, clothing, and maintenance." Henry S. Richardson, *Greenwich: Its History, Antiquities, Improvements, and Public Buildings* (Greenwich: Harriet Richardson, 1834), 42. For context, 1 s was thirty times less than a weekly wage that would have provided "a very comfortable working-class income" Simon Eliot, "The Business of Victorian Publishing," in *The Cambridge Companion to the Victorian Novel*, edited by Deirdre David (Cambridge: Cambridge University Press, 2000), 37.
14. John Morreall, "Humour and the Conduct of Politics," in *Beyond a Joke: The Limits of Humour*, ed. Michael Pickering and Sharon Lockyer (Basingstoke: Palgrave, 2005), 66. The incongruity model differs from the superiority model, which dominated from the time of Plato to the eighteenth century. The superiority model suggested that humour emanates from a feeling of superiority over either someone else or a former state of ourselves. A third model for understanding how humour functions is the relief model, which suggests that laughter is the release of energy in the nervous system that has been suddenly rendered unneeded (Morreall, "Humour," 65–68). Morreall shows how the release model overlaps with the incongruity model ("Humour," 67–68).
15. Stephen Leacock, *Humor: Its Theory and Technique, with Examples and Samples; a Book of Discovery* (New York: Dodd, Mead, 1935), 15.
16. Mitchell and Snyder, *Narrative Prosthesis*; Tom Coogan and Rebecca Mallett, "Introduction: Disability, Humour and Comedy," *Journal of Literary & Cultural Disability Studies* 7, no. 3 (2013).
17. Coogan and Mallett, "Introduction," 68.

18. Morreall, "Humour," 67–68.
19. Tom Shakespeare, "Joking a Part," *Body & Society* 5, no. 4 (1999): 49.
20. For more on the rise of this genre, see Diana Donald, *The Age of Caricature: Satirical Prints in the Reign of George III* (New Haven: Yale University Press, 1996).
21. Jonathan Swift, "A Beautiful Young Nymph Going to Be," Poetry Foundation, accessed July 2, 2018, https://www.poetryfoundation.org/poems/50580/a-beautiful-young-nymph-going-to-bed.
22. Simon Dickie, "Hilarity and Pitilessness in the Mid-Eighteenth Century: English Jestbook Humor," *Eighteenth-Century Studies* 37, no. 1 (2003). Elsewhere, as Amelia F. Rauser has shown us, caricatures focusing on the excessiveness of wigs were regularly used to criticize macaronis. Amelia F. Rauser, "Hair, Authenticity, and the Self-Made Macaroni," *Eighteenth-Century Studies* 38, no. 1 (2004).
23. Chase, *The Victorians*, 276.
24. John Woodforde, *The Strange Story of False Teeth* (London: Routledge and Kegan Paul, 1968), 87–92.
25. Galia Ofek, *Representations of Hair in Victorian Literature and Culture* (Aldershot: Ashgate, 2009), 28.
26. Woodforde, *The Strange*, 2.
27. Edgar Allan Poe, "The Spectacles," in *Prose Tales (Second Series) Arthur Gordon Pym*, by Edgard Allan Poe (Boston: Dana Estes, 1884).
28. Poe, "The Man," 315.
29. Poe, "The Spectacles," 164, 195, 194, and 172.
30. E. T. A. Hoffman, "The Sandman," in *Tales from the German, Comprising Specimens from the Most Celebrated Authors*, trans. John Oxenford and C. A. Feiling (London: Chapman and Hall 1844), Project Gutenberg, last modified April 18, 2010, https://www.gutenberg.org/files/32046/32046-h/32046-h.htm.
31. Sharrona Pearl, *About Faces: Physiognomy in Nineteenth-Century Britain* (Cambridge: Harvard University Press, 2010), 221.
32. Martha Stoddard Holmes and Mark Mossman, "Disability in Victorian Sensation Fiction," in *A Companion to Sensation Fiction*, ed. Pamela K. Gilbert (Chichester: Wiley-Blackwell, 2011), 493.
33. Wilkie Collins, *The Black Robe* (London: Chatto and Windus, 1910).
34. Lisa Niles, "Owning 'the Dreadful Truth'; or, Is Thirty-Five Too Old?: Age and the Marriageable Body in Wilkie Collins's *Armadale*," *Nineteenth-Century Literature* 65, no. 1 (2010): 68.
35. Niles, "Owning," 77–79.
36. Qtd. in ibid., 78.
37. Niles, "Owning," 79.

38. Collins, *The Black*, 67–68.
39. William Makepeace Thackeray, *The Virginians*, 2 vols. (London: Bradbury and Evans, 1858–1859); Anthony Trollope, *The Way We Live Now* (London: Penguin, 1994).
40. Collins, *The Black*, 67.
41. Margaret Morganroth Gullette, *Aged by Culture* (Chicago: University of Chicago Press, 2004), 161–62.
42. Collins, *The Black*, 67–68.
43. Susan Sontag, "The Double Standard of Aging," in Psychology of Women: Behavior in a Biosocial Context, ed. Juanita H. Williams, 3rd ed. (London: Norton, 1987).
44. Collins, *The Law*, 59–60.
45. Ibid., 62.
46. Ibid., 190.
47. Ibid., 64.
48. Ibid., 59.
49. Ofek, *Representations*, 37; Collins, *The Law*, 189, 263, and 408.
50. John Woodforde, *The Strange Story of False Hair* (London: Routledge and Kegan Paul, 1971), 76.
51. "The New Wig Club," *Punch*, May 26, 1855, 209.
52. Ibid.
53. "The Old Man and His Grey Wig." *The Age*, February 12, 1832.
54. Ibid.
55. Collins, *The Law*, 409.
56. Ibid., 409–10.
57. Ibid., 408.
58. Ibid., 257.
59. Alexander Rowland, *The Human Hair, Popularly and Physiologically Considered with Special Reference to Its Preservation, Improvement and Adornment, and the Various Modes of Its Decoration in All Countries* (London: Piper, Brothers & Co., 1853), 159. Thomas Elliott, "Hair Cutter, Hair Grower, Wig Maker, and Perfumer," of Fenchurch Street, London, advertised his "Feather-weight wigs" for £1 10s circa 1860, verifying Rowland's estimate. Thomas Elliott, Hair Cutter, Hair Grower, Wig Maker, and Perfumer, advertisement, [c.1860], EPH 160B, hair care box 9, Wellcome Library, London.
60. Gregory Clark, "What Were the British Earnings and Prices Then? (New Series)," Measuring Worth, accessed July 2, 2018, http://www.measuringworth.com/ukearncpi/.
61. Collins, *Armadale*, 236–37.
62. Collins, *Armadale*, 279.

63. Ibid.
64. Ibid., 454.
65. As Woodforde notes, "[toothbrushes] were rarities, or at any rate luxuries, till after about 1850." Chandra and Chandra suggest that the term "oral hygiene" was not even coined until 1884. Woodforde, *The Strange Story of False Teeth*, 45; Satish Chandra and Shaleen Chandra, *Textbook of Community Dentistry* (New Delhi: Jaypee Brothers Publishers, 2002), 7.
66. Woodforde, *The Strange Story of False Teeth*, 43.
67. Varney is described at one point as having an "almost beautiful smile, which displayed his white glistening teeth to perfection." [James Malcom Rymer], *Varney the Vampire; or, The Feast of Blood*, Project Gutenberg, last modified October 4, 2011, http://www.gutenberg.org/files/14833/14833-h/14833-h.htm, ch. 8.
68. In Chapter 34 of Brontë's novel, Heathcliff is described as having "sharp white teeth." Emily Brontë, *Wuthering Heights* (London: John Murray, 1910), Project Gutenberg, last modified August 28, 2010, https://www.gutenberg.org/files/768/768-h/768-h.htm. For more on the way that Brontë evokes debate concerning Heathcliff's possible status as a vampire, see Carol A. Senf, *The Vampire in Nineteenth Century English Literature* (Madison: University of Wisconsin Press, 2013), 75–93.
69. Collins, *Armadale*, 566.
70. Bill Brown, "Thing Theory," *Critical Inquiry* 28, no. 1 (2001): 4.
71. Mark Hampton, *Visions of the Press in Britain, 1850–1950* (Urbana: University of Illinois Press, 2004), 119.
72. Nicholson, "Jonathan's Jokes," 37.
73. See Mr. H. Lyon and Mr. Tindall, Dentists, advertisement, *Ipswich Journal*, January 2, 1878. Before the introduction of vulcanite, gold and ivory were the most popular materials used for making denture bases (Woodforde, *The Strange Story of False Teeth*, 87–88).
74. Grace Goldney, *Marion's Choice*, *Aunt Judy's Magazine* 8, no. 112–14 (1870); William Henry Archibald Chasemore, "Wicklebury's Wig," *Judy*, January 2, 1878; "Mr. Sagtooth's First Appearance," *Funny Folks* 17, no. 853 (1891); Harold Copping, "It Caught His Wig and Jerked It into the Air," wood-engraved illustration, in *Hazelbrake Hollow*, by F. Scarlett Potter, *The Children's Friend* (1895).
75. C. D. Shanly, "The Hair, the Moustache, and the Beard," *The Treasury of Literature and the Ladies' Treasury*, February 1, 1872, 84.
76. Woodforde, *The Strange Story of False Hair*, 78.
77. "To Wearers of False Hair." *The Age*, May 1, 1831, 139.
78. G. Renaud, "Waggles's Wig," *Judy; or The London Serio-Comic Journal*, March 9, 1887.

79. "The Story of Gaffer Grey," *Little Folks*, no. 45–48 (1871); C. E. Pearce, "Princess Prettypet," *Our Young Folks Weekly Budget of Tales, News, Sketches, Fun, Puzzles, Riddles &c.*, November 22, 1873; "The General's Wig," *Little Wide-Awake*, [1875].
80. *The Good-Natured Boy; or, What Came of It. Boys of England: A Journal of Sport, Travel, Fun and Instruction for the Youths of All Nations* 28–9, no. 719–36 (1880).
81. Mark Twain, "Man's Place in the Animal World," in *Mark Twain's Book of Animals*, ed. Shelley Fisher Fishkin (Berkeley: University of California Press, 2009), 124.
82. Mark Twain, *Roughing It* (London: Penguin, 1988), 386–87.
83. Alfred Edersheim, *Shorn to the Wind*, *The Sunday at Home: A Family Magazine for Sabbath Reading*, no. 1014–26 (1873).
84. "None So Blind as Those Who Won't See." *Standard*, October 5, 1846; "An Injured Frenchwoman." *Bristol Mercury*, May 17, 1862; "A Glass Eye Expected to See!" *Bow Bells* 6, no. 131 (1867).
85. "A Glass Eye Expected to See!" 4.
86. "An' they tell me that you've got a glass-eye," *Preston Guardian*, July 19, 1884, 2.
87. D. B. McKean, "A Wig and a Wooden Leg," *Chatterbox*, no. 4 (1886): 27; "A Cure for Cannibalism." *Pick-Me-Up*, no. 37 (1889): 171.
88. "A Cure," 171.
89. Richard F. Patteson, "'King Solomon's Mines': Imperialism and Narrative Structure," *The Journal of Narrative Technique* 8, no. 2 (1978): 113; Joseph A. Kestner, *Masculinities in British Adventure Fiction, 1880–1915* (Burlington: Ashgate, 2010), 67.
90. Kay Heath, *Aging by the Book: The Emergence of Midlife in Victorian Britain* (Albany: State University of New York Press, 2009), 2.
91. Ibid., 13.
92. Haggard, *King*, ch. 1. The narrator and protagonist of Haggard's novel, Quatermain, suggests that Good's release from the navy came because "it was impossible that he should be promoted" (ch. 1), alluding, possibly, to the fact that the Captain was not from a prestigious enough family to be promoted further. But such a conclusion does not really explain why he was let go. His deficiencies seem to better explain this decision. Despite his reputation as a "marvellous shot"—established after shooting a moving giraffe calf from 300 yards with his last shot—Quatermain reveals that his companion was in fact a bad shot, hinting that Good may be past his best in terms of military prowess.
93. McKean, "A Wig," 27.
94. "A Cure," 171.

95. Haggard, *King*, ch. 1.
96. Ibid.
97. Ibid., ch. 7.
98. Ibid., ch. 1.
99. Poe, "The Man."
100. Bruce Haley, *The Healthy Body and Victorian Culture* (Cambridge: Harvard University Press, 1978); Erin O'Connor, *Raw Material: Producing Pathology in Victorian Culture* (Durham: Duke University Press, 2000).
101. Heath, *Aging*, 13 and 29.
102. Norman Vance, *The Sinews of the Spirit: The Ideal of Christian Manliness in Victorian Literature and Religious Thought* (Cambridge: Cambridge University Press, 1985), 166; Heath, *Aging*, 29.
103. Heath, *Aging*, 28.
104. John Tosh, *Manliness and Masculinities in Nineteenth-Century Britain: Essays on Gender, Family, and Empire* (Harlow: Pearson Longman, 2005).
105. Haggard, *King*, ch. 7.
106. Helen Small has shown us that many of these stereotypes have existed since Aristotelean times. Helen Small, *The Long Life* (Oxford: Oxford University Press, 2007), 52–88.
107. Good's false teeth have been read in various compelling ways, but few have considered the device's comic purpose and how this relates to ageing. Joseph Kestner sees Good's use of dentures as evidence for what Richard F. Patteson's identifies as a key imperial trope: how "heroes establish their influence over natives through a technological device." Merrick Burrow, on the other hand, suggests that Good's teeth are emblematic of "metropolitan artifice," which are perceived as "sacred relics" by the Kukuanaland people. The white men are forced to accept such a position to survive thereby valorizing a masculine adoption of barbarism (a stance that aligns Burrow's work with that of Bradley Deane). Deane observes that "Good's fastidiousness begins to make him a figure of fun," but that is as close as we get in Haggard scholarship to an acknowledgement that Good's teeth are a comic prop representative of a widely used humorous trope. Kestner, *Masculinities*, 67; Patteson, "King," 113; Merrick Burrow, "The Imperial Souvenir: Things and Masculinities in H. Rider Haggard's King Solomon's Mines and Allan Quatermain," *Journal of Victorian Culture* 18, no. 1 (2013): 85; Bradley Deane, "Imperial Barbarians: Primitive Masculinity in Lost World Fiction," *Victorian Literature and Culture* 36, no. 1 (2008): 221n11.
108. Haggard, *King*, ch. 4.

109. Robert Louis Stevenson, *Treasure Island* (New York: Harper and Brothers, 1915). As Mavis Reimer writes, "Stevenson's novel prompted H. Rider Haggard to define his novel, *King Solomon's Mines* (1885), as intended for 'all the big and little boys who read', a readership he saw as distinguishing his work from that of 'people who write books for little girls in the school-room.'" Mavis Reimer, "'These Two Irreconcilable Things—Art and Young Girls': The Case of the Girls' School Story," in *Girls, Boys, Books, Toys: Gender in Children's Literature and Culture*, eds. Beverly Lyon Clark and Margaret R. Higonnet (Baltimore: John Hopkins University Press, 2000), 44.
110. Georgia Haire, "'Does It Really Make You Look Old?': Dentures, Ageing and Aesthetics in Late Twentieth-Century Britain," workshop paper presented at "Oral Health Inequalities, Oral Hygiene Cultures: Past, Present, Future," University of Kent, June 28, 2018.

References

Bigg, Henry Robert Heather. *Artificial Limbs, and the Amputations Which Afford the Most Appropriate Stumps in Civil and Military Surgery*. London: Henry Robert Heather Bigg, 1885.

Boyer, George R., and Timothy P. Schmidle. "Poverty among the Elderly in Late Victorian England." *The Economic History Review* 62, no. 2 (2009): 249–78.

"An Injured Frenchwoman". *The Bristol Mercury*, May 17, 1862.

Brontë, Emily. *Wuthering Heights*. London: John Murray, 1910. Project Gutenberg. Last modified August 28, 2010. https://www.gutenberg.org/files/768/768-h/768-h.htm.

Brown, Bill. "Thing Theory." *Critical Inquiry* 28, no. 1 (2001): 1–22.

Burrow, Merrick. "The Imperial Souvenir: Things and Masculinities in H. Rider Haggard's King Solomon's Mines and Allan Quatermain." *Journal of Victorian Culture* 18, no. 1 (2013): 72–92.

Chandra, Satish, and Shaleen Chandra. *Textbook of Community Dentistry*. New Delhi: Jaypee Brothers Publishers, 2002.

Chase, Karen. *The Victorians and Old Age*. Oxford: Oxford University Press, 2009.

———. "'Senile' Sexuality." In *Interdisciplinary Perspectives on Aging in Nineteenth-Century Culture*, edited by Katharina Boehm, Anna Farkas, and Anne-Julia Zwierlein, 132–46. New York: Routledge, 2014.

Chasemore, William Henry Archibald. "Wicklebury's Wig." *Judy*, January 2, 1878.

Collins, Wilkie. *Armadale*. Oxford: Oxford University Press, 1999.

———. *The Law and the Lady*. Oxford: Oxford University Press, 2008.

———. *The Black Robe*. London: Chatto and Windus, 1910.

Coogan, Tom, and Rebecca Mallett. "Introduction: Disability, Humour and Comedy." *Journal of Literary & Cultural Disability Studies* 7, no. 3 (2013): 247–53.

Copping, Harold. "It Caught His Wig and Jerked It into the Air." Wood-engraved illustration. In *Hazelbrake Hollow*, by F. Scarlett Potter. *The Children's Friend* (1895): n.pag.

"A Cure for Cannibalism". *Pick-Me-Up*, no. 37 (1889): 171.

Deane, Bradley. "Imperial Barbarians: Primitive Masculinity in Lost World Fiction." *Victorian Literature and Culture* 36, no. 1 (2008): 205–25.

Dickens, Charles. *Dombey and Son*. London: Penguin, 2002.

———. *Our Mutual Friend*. London: Everyman, 2000.

Dickie, Simon. "Hilarity and Pitilessness in the Mid-Eighteenth Century: English Jestbook Humor." *Eighteenth-Century Studies* 37, no. 1 (2003): 1–22.

Donald, Diana. *The Age of Caricature: Satirical Prints in the Reign of George III*. New Haven: Yale University Press, 1996.

Edersheim, Alfred. *Shorn to the Wind. The Sunday at Home: A Family Magazine for Sabbath Reading*, no. 1014–26 (1873): 625–28; 641–45; 657–61; 673–77; 689–93; 705–09; 721–26; 737–40; 753–56; 769–72; 785–89; 801–05; 817–22.

Eliot, Simon. "The Business of Victorian Publishing." In *The Cambridge Companion to the Victorian Novel*, edited by Deirdre David, 37–61. Cambridge: Cambridge University Press, 2000.

"A Strange Complaint". *Fun*, February 18, 1871.

"The General's Wig". *Little Wide-Awake*, [1875].

"A Glass Eye Expected to See!" *Bow Bells* 6, no. 131 (1867): 4.

Goldney, Grace. *Marion's Choice. Aunt Judy's Magazine* 8, no. 112–14 (1870): n. pag.

The Good-Natured Boy; or, What Came of It. Boys of England: A Journal of Sport, Travel, Fun and Instruction for the Youths of All Nations 28–9, no. 719–36 (1880): 281–83, 313–15, 326–28, 342–44, 374–76; 6–8, 38–40, 54–56, 70–72, 102–4, 118–20, 143–45.

Gullette, Margaret Morganroth. *Aged by Culture*. Chicago: University of Chicago Press, 2004.

Haggard, H. Rider. *King Solomon's Mines*. Project Gutenberg. Last modified October 12, 2012. https://www.gutenberg.org/files/2166/2166-h/2166-h.htm.

Haire, Georgia. "'Does It Really Make You Look Old?': Dentures, Ageing and Aesthetics in Late Twentieth-Century Britain." Workshop paper presented at "Oral Health Inequalities, Oral Hygiene Cultures: Past, Present, Future," University of Kent, June 28, 2018.

Haley, Bruce. *The Healthy Body and Victorian Culture*. Cambridge: Harvard University Press, 1978.

Hampton, Mark. *Visions of the Press in Britain, 1850–1950*. Urbana: University of Illinois Press, 2004.

Heath, Kay. *Aging by the Book: The Emergence of Midlife in Victorian Britain*. Albany: State University of New York Press, 2009.

Hoffman, E. T. A. "The Sandman." In *Tales from the German, Comprising Specimens from the Most Celebrated Authors*, translated by John Oxenford and C. A. Feiling, n.pag. London: Chapman and Hall 1844. Project Gutenberg. Last modified April 18, 2010. https://www.gutenberg.org/files/32046/32046-h/32046-h.htm.

Holmes, Martha Stoddard, and Mark Mossman. "Disability in Victorian Sensation Fiction." In *A Companion to Sensation Fiction*, edited by Pamela K. Gilbert, 493–506. Chichester: Wiley-Blackwell, 2011.

[Horne, Richard H.] "Dust; or Ugliness Redeemed." *Household Words* 1, no. 16 (1850): 379–84.

Kestner, Joseph A. *Masculinities in British Adventure Fiction, 1880–1915*. Burlington: Ashgate, 2010.

"The Rewards of Warfare". *Kind Words: A Week-Day Magazine for Boys & Girls*, November 24, 1870.

"Laid up at Greenwich". *Judy: or The London Serio-Comic Journal*, December 2, 1868.

Leacock, Stephen. *Humor: Its Theory and Technique, with Examples and Samples; a Book of Discovery*. New York: Dodd, Mead, 1935.

Messrs Mosley's Teeth and Painless Dentistry. Advertisement. *Bell's Life in London and Sporting Chronicle* 2, no. 365 (1867): 3.

Mitchell, David T., and Sharon L. Snyder. *Narrative Prosthesis: Disability and the Dependencies of Discourse*. Ann Arbor: University of Michigan Press, 2000.

"Modern Sonnet. To an Old Wig". In *The Spirit of the Public Journals for 1802*, vol. 6, 112–13. London: James Ridgway, 1803.

Morreall, John. "Humour and the Conduct of Politics." In *Beyond a Joke: The Limits of Humour*, edited by Michael Pickering and Sharon Lockyer, 63–78. Basingstoke: Palgrave, 2005.

Mr. H. Lyon and Mr. Tindall, Dentists. Advertisement. *Ipswich Journal*, January 2, 1878.

"Mr. Sagtooth's First Appearance". *Funny Folks* 17, no. 853 (1891): 108–109.

Mr. Scott, Surgeon and Mechanical Dentist. Advertisement. *Bell's Life in London and Sporting Chronicle* 417 (1830): 2.

"Mr. Wegg and His Class". *All the Year Round* 18, no. 441 (1877): 250–53.

Multiple classified advertisements. *Lady's Newspaper*, no. 599 (1858): 399.

"The New Wig Club". *Punch*, May 26, 1855.

Nicholson, Bob. "Jonathan's Jokes." *Media History* 18, no. 1 (2012): 33–49.

Niles, Lisa. "Owning 'the Dreadful Truth'; or, Is Thirty-Five Too Old?: Age and the Marriageable Body in Wilkie Collins's *Armadale*." *Nineteenth-Century Literature* 65, no. 1 (2010): 65–92.

O'Connor, Erin. *Raw Material: Producing Pathology in Victorian Culture*. Durham: Duke University Press, 2000.

Ofek, Galia. *Representations of Hair in Victorian Literature and Culture*. Aldershot: Ashgate, 2009.

"The Old Man and His Grey Wig". *The Age*, February 12, 1832.

Patteson, Richard F. "'King Solomon's Mines': Imperialism and Narrative Structure." *The Journal of Narrative Technique* 8, no. 2 (1978): 112–23.

Pearce, C. E. "Princess Prettypet." *Our Young Folks Weekly Budget of Tales, News, Sketches, Fun, Puzzles, Riddles &c.*, November 22, 1873.

Pearl, Sharrona. *About Faces: Physiognomy in Nineteenth-Century Britain*. Cambridge: Harvard University Press, 2010.

Poe, Edgar Allan. "The Man That Was Used Up: A Tale of the Late Bugaboo and Kickapoo Campaign." In *The Works of the Late Edgar Allan Poe*, by Edgar Allan Poe, vol. 4, 315–25. New York: Blakeman & Mason, 1859.

———. "The Spectacles." In *Prose Tales (Second Series) Arthur Gordon Pym*, by Edgard Allan Poe, 161–98. Boston: Dana Estes, 1884.

"An' they tell me that you've got a glass-eye". *Preston Guardian*, July 19, 1884.

Rauser, Amelia F. "Hair, Authenticity, and the Self-Made Macaroni." *Eighteenth-Century Studies* 38, no. 1 (2004): 101–17.

Reimer, Mavis. "'These Two Irreconcilable Things—Art and Young Girls': The Case of the Girls' School Story." In *Girls, Boys, Books, Toys: Gender in Children's Literature and Culture*, edited by Beverly Lyon Clark and Margaret R. Higonnet, 40–52. Baltimore: John Hopkins University Press, 2000.

Renaud, G. "Waggles's Wig." *Judy; or The London Serio-Comic Journal*, March 9, 1887.

Richardson, Henry S. *Greenwich: Its History, Antiquities, Improvements, and Public Buildings*. Greenwich: Harriet Richardson, 1834.

Rowland, Alexander. *The Human Hair, Popularly and Physiologically Considered with Special Reference to Its Preservation, Improvement and Adornment, and the Various Modes of Its Decoration in All Countries*. London: Piper, Brothers & Co., 1853.

[Rymer, James Malcom]. *Varney the Vampire; or, The Feast of Blood*. Project Gutenberg. Last modified October 4, 2011. http://www.gutenberg.org/files/14833/14833-h/14833-h.htm.

Senf, Carol A. *The Vampire in Nineteenth Century English Literature*. Madison: University of Wisconsin Press, 2013.

Shakespeare, Tom. "Joking a Part." *Body & Society* 5, no. 4 (1999): 47–52.

Shanly, C. D. "The Hair, the Moustache, and the Beard." *The Treasury of Literature and the Ladies' Treasury*, February 1, 1872.

Small, Helen. *The Long Life*. Oxford: Oxford University Press, 2007.

Sontag, Susan. "The Double Standard of Aging." In *Psychology of Women: Behavior in a Biosocial Context*, edited by Juanita H. Williams, 462–78. 3rd ed. London: Norton, 1987.

"None So Blind as Those Who Won't See". *The Standard*, October 5, 1846.

Stevenson, Robert Louis. *Treasure Island*. New York: Harper and Brothers, 1915.

"The Story of Gaffer Grey". *Little Folks*, no. 45–48 (1871): 292–93; 314–15; 322–23; 338–39.

Swift, Jonathan. "A Beautiful Young Nymph Going to Bed." Poetry Foundation. Accessed July 2, 2018. https://www.poetryfoundation.org/poems/50580/a-beautiful-young-nymph-going-to-bed.

Thackeray, William Makepeace. *The Virginians*. 2 vols. London: Bradbury and Evans, 1858–1859.

Thane, Pat. *Old Age in English History: Past Experiences, Present Issues*. Oxford: Oxford University Press, 2002.

———. "Epilogue." In *Interdisciplinary Perspectives on Aging in Nineteenth-Century Culture*, edited by Anne-Julia Zwierlein, Katharina Boehm, and Anna Farkas, 232–46. New York: Routledge, 2014.

Thomas Elliott, Hair Cutter, Hair Grower, Wig Maker, and Perfumer. Advertisement. [c.1860]. EPH 160B. Hair care box 9. Wellcome Library, London.

"Thrift, or Nothing Is Useless". *Chambers's Edinburgh Journal* 1, no. 38 (1844): 191–92.

Tosh, John. *Manliness and Masculinities in Nineteenth-Century Britain: Essays on Gender, Family, and Empire*. Harlow: Pearson Longman, 2005.

"To Wearers of False Hair". *The Age*, May 1, 1831.

Trollope, Anthony. *The Way We Live Now*. London: Penguin, 1994.

Twain, Mark. *Roughing It*. London: Penguin, 1988.

———. "Man's Place in the Animal World." In *Mark Twain's Book of Animals*, edited by Shelley Fisher Fishkin, 117–25. Berkeley: University of California Press, 2009.

Vance, Norman. *The Sinews of the Spirit: The Ideal of Christian Manliness in Victorian Literature and Religious Thought*. Cambridge: Cambridge University Press, 1985.

Woodforde, John. *The Strange Story of False Teeth*. London: Routledge and Kegan Paul, 1968.

———. *The Strange Story of False Hair*. London: Routledge and Kegan Paul, 1971.

Zwierlein, Anne-Julia, Katharina Boehm, and Anna Farkas, eds. *Interdisciplinary Perspectives on Aging in Nineteenth-Century Culture*. New York: Routledge, 2014.

CHAPTER 7

Conclusion

Disabled people don't have to play the villain. The Superhumans Return.
—Channel 4, Advertisement for Channel 4's coverage of the Paralympic Games 2016, YouTube, last modified, November 6, 2015, *https://www.youtube.com/watch?v=1FKcmEXPnUw.*

"The Superhumans Return"—a provocative advertisement for the 2016 Paralympic Games by the British free-to-air television broadcasting network Channel 4—intentionally draws attention to and problematizes the still-dominant trope of the disabled villain. Still, it also unwittingly reifies another problematic stereotype: the "supercrip."[1] The ad depicts five subjects: two artificial-limb users, a wheelchair user, a person with a congenital deformity of the arm, and a person of small stature—each of whom plays the role of a stereotypical melodramatic villain. This critical stance aligns the ad with the work of disability-studies scholars and activists such as Paul K. Longmore, Jenny Morris, David T. Mitchell and Sharon L. Snyder, Ato Quayson, and David Roche, who critique the cultural trope of physical difference as a metaphor for evil and/or moral corruption.[2] Bringing a critical attitude to disability representation into popular consciousness is certainly encouraging. But in labelling Paralympic athletes "superhumans," and by extending its advertising campaign for the

R. Sweet, *Prosthetic Body Parts in Nineteenth-Century Literature and Culture*, Palgrave Studies in Nineteenth-Century Writing and Culture, https://doi.org/10.1007/978-3-030-78589-5_7

2012 Paralympic Games that used the same contentious branding, Channel 4 falls into a trap of undermining the complex humanity of Paralympic athletes through an overemphasis on their "overcoming-the-odds" backstories. This label presents disabled athletes as more than human, spectacular, even freakish. The *super* in "superhuman" derives from Latin, where the term is used chiefly with the sense "above, over" (of place) (*Oxford English Dictionary*).

It is certainly encouraging that athletes living with physical impairments are provided a popular and engagingly promoted public platform upon which to gain recognition for their achievements. Nonetheless, the very term "superhuman" weakens such a project by reinforcing the ableist tone that often inflects discussions about disabled sportspersons: "it's amazing what s/he has achieved given *X*." As Carla Filomena Silva and P. David Howe explain, "supercrip narratives may have a negative impact on the physical and social development of disabled individuals by reinforcing what could be termed 'achievement syndrome'—the impaired are successful in spite of their disability."[3] In an informative companion video (available on both Channel 4 website and YouTube), which accompanies the "Superhumans Return" advert, the disabled actors that star in the ad provide comments on current attitudes to physical difference, drawing in particular on their experience of applying for acting roles in film and television.[4] The actors reveal how they often struggle to get roles that do not programmatically exploit their physical differences for narrative purposes. The film ends provocatively with one female actor stating, "Just because I'm 4 foot 6 and you're not. We're just still human beings. We're still connected."[5] Unfortunately, this empathetic statement is not matched by the ad itself, which replaces one limiting stereotype (the disabled villain) with another (the supercrip).

Clearly, representations of physical difference have moved on in various ways since the period that I have explored in *Prosthetic Body Parts in Nineteenth-Century Literature and Culture*. But there remain overlaps between then and now. For instance, the trope of the prosthesis-using villain is evoked, a representational typecast that we have seen has deep historical roots, notwithstanding Victorian precursors such as Silas Wegg from Charles Dickens's *Our Mutual Friend* (1864–1865) and Jonathan Small from Arthur Conan Doyle's Sherlock Holmes novel *The Sign of Four* (1890).[6] Similarly, the motif of the weaponized prosthetic body part is also redeployed. Reminiscent of Robocop, one male artificial leg user is shown to have a cybernetic prosthesis that incorporates a holster for a pistol. As shown in Chap. 3, this icon remains popular in cultural depictions of prosthetic body parts, particularly as we move ever closer what some have called a "transhuman" age. We live in a time now when advertisements including "The Superhumans Return" are released to the public with the intent of challenging assumptions about what it means to be disabled or nondisabled. The physically different, often prostheticized body, is presented to us as a viable alternative to organic physical wholeness, which, though increasingly challenged, remains hegemonic. However, advances in the field of disability studies have placed the dominance of physical completeness under much-needed critical scrutiny.

This book has argued that narratives from the very period in which prostheses saw their most significant technological changes throw into question the cultural privileging of physical wholeness. Does such questioning suggest that the nineteenth century marked an era of progressive enlightenment? As my case studies show, representational tropes, including that of the beggar with wooden leg (explored in depth in Chap. 4) or the failing cosmetic prosthesis (investigated in detail in Chaps. 5 and 6), endured throughout the period. Specific historical and cultural factors, such as the 1860s fashion for false hair, informed specific manifestations of prosthesis tropes and yet many of the representations drew from previous depictions and often questioned the dominance of physical wholeness. What is true, however, is that the number of cultural and literary representations spiked around periods when these devices saw major innovation or increased circulation. For instance, even in Britain the 1860s saw an increase in mentions of the term "artificial leg" according to an "entire document" search of the ProQuest source *British Periodicals Online* (Collections I, II, and III).[7] This spike correlates with the developments in prosthesis manufacture and distribution to American Civil War amputees. A comparable search of *British Periodicals Online*, using the term "artificial eye," also yields a peak in the 1860s.[8] This increase can be attributed to the presence in England of Parisian artificial-eye maker Auguste Boissonneau, whose enamel artificial eyes dominated the European market in the 1840s, 1850s, and 1860s.

Regarding the specific ways that the hegemony of wholeness was challenged by the prosthesis narratives, we have seen a variety of approaches. Chapter 3 revealed how representations of highly effective prostheses, devices that could be used as weapons, and self-acting prosthetics provided challenges to the hegemonic concept of physical wholeness by presenting menacingly powerful and at times intelligent devices—non-human parts that threatened to usurp the organic whole. Chapter 4 explored prosthesis users who threatened to upset the cultural applecart by advancing their social positions. What was most transgressive about the stories explored in this chapter was the fact that the success that several of the prosthesis users achieved stemmed precisely from the conspicuousness of their artificial body parts rather than from their ability to enable their users to pass. Chapter 5 similarly showed how prostheses were sometimes imagined as desirable assets for a wide variety of reasons in nineteenth-century marriage plots. Paradoxes inherent in the social system that privileged wholeness were interrogated, as Chap. 6 has explained, by narratives that humorously depicted ageing prosthesis users. For many Victorian writers, the privileging of wholeness had brought about an army of aged prosthesis users, who provided substantial material for comic representations.

In conclusion, *Prosthetic Body Parts in Nineteenth-Century Literature and Culture* adds to our understanding of the history of disability, the construction of normalcy, and the relationship between literature and science, technology, and medicine. The prosthetic has been popular as both a critical metaphor and a material artefact for scholarly inquiry during the past twenty years, but there remains much to be understood, especially about the longer literary history concerning this technology. My research provides part of the story, though several angles remain uncovered. How were literary representations of prosthesis affected by the First World War, the Second World War, and artistic movements of the twentieth century such as modernism? How does race affect representations of prosthesis users? How were prostheses treated in non-Anglophone literatures? Humans are not the only species to use prostheses. Today, a wide variety of non-human animals are fitted with prosthetic body parts for both cosmetic and compassionate reasons. What can be said about the cultural history of this phenomenon? David Wills brings to the fore the ontological complexity of the prosthetic when he writes: "[T]he writing of prosthesis ... is inevitably caught in a complex play of displacements; prosthesis being about nothing if not placement, displacement, replacement, standing, dislodging, substituting, setting, amputating, supplementing."[9] I hope to have shown how the intricacy of the prosthetic is matched by its remarkable nineteenth-century cultural and literary history.

Notes

1. R. J. Berger describes supercrips as "those individuals whose inspirational stories of courage, dedication, and hard work prove that it can be done, that one can defy the odds and accomplish the impossible." For Marie Myers Hardin and Brent Hardin, such a model involves presenting the disabled person as heroic due to his or her ability to perform feats normally considered impossible for people with disabilities or by virtue of the person living a "regular" life despite impairment. In Carla Filomena Silva and P. David Howe's article, supercrip "implies a stereotyping process that requires an individual 'to fight against his/her impairment' in order to overcome it and achieve unlikely 'success.'" R. J. Berger, "Disability and the Dedicated Wheelchair Athlete: Beyond the 'Supercrip' Critique," *Journal of Contemporary Ethnography* 37, no. 6 (2008): 648; Marie Myers Hardin and Brent Hardin, "The 'Supercrip' in Sport Media: Wheelchair Athletes Discuss Hegemony's Disabled Hero," *Sociology of Sport Online* 7, no. 1 (2004): 5.3; Carla Filomena Silva and P. David Howe, "The (In)validity of Supercrip Representation of Paralympian Athletes," *Journal of Sport and Social Issues* 36, no. 2 (2012): 175.
2. Paul K. Longmore, "Screening Stereotypes: Images of Disabled People in Television and Motion Pictures," in *Images of the Disabled, Disabling Images*, ed. A. Ian Gartner (New York: Praeger, 1987); Jenny Morris, *Pride against Prejudice: Transforming Attitudes to Disability* (London: The Women's Press, 1996); David T. Mitchell and Sharon L. Snyder, *Narrative Prosthesis: Disability and the Dependencies of Discourse* (Ann Arbor: University of Michigan Press, 2000); Ato Quayson, *Aesthetic Nervousness: Disability and the Crisis of Representation* (New York: Columbia University Press, 2007); David Roche, "The Metaphor of Facial Disfigurement," *Huffington Post*, May 25, 2011, http://www.huffingtonpost.com/david-roche/the-metaphor-of-facial-di_b_144949.html.
3. Silva and Howe, "The (In)validity," 174.
4. Channel 4, "Meet the Cast: The Superhumans Return," YouTube, last modified November 6, 2015, https://www.youtube.com/watch?v=cGnHzFldPwY.
5. Ibid.
6. Charles Dickens, *Our Mutual Friend* (London: Everyman, 2000); Arthur Conan Doyle, *The Sign of Four* (London: Penguin, 2001).
7. ProQuest, *British Periodicals*, accessed July 5, 2018, https://www.proquest.com/products-services/british_periodicals.html.
8. Sixteen results are produced as opposed to seven in 1850–1859 and four in 1870–1879.
9. David Wills, *Prosthesis* (Stanford: Stanford University Press, 1995), 9.

References

Berger, R. J. "Disability and the Dedicated Wheelchair Athlete: Beyond the 'Supercrip' Critique." *Journal of Contemporary Ethnography* 37, no. 6 (2008): 647–78.

Channel 4. Advertisement for Channel 4's coverage of the Paralympic Games 2016. YouTube. Last modified November 6, 2015. https://www.youtube.com/watch?v=1FKcmEXPnUw.

———. "Meet the Cast: The Superhumans Return." YouTube. Last modified November 6, 2015. https://www.youtube.com/watch?v=cGnHzFldPwY.

Dickens, Charles. *Our Mutual Friend.* London: Everyman, 2000.

Doyle, Arthur Conan. *The Sign of Four.* London: Penguin, 2001.

Hardin, Marie Myers, and Brent Hardin. "The 'Supercrip' in Sport Media: Wheelchair Athletes Discuss Hegemony's Disabled Hero." *Sociology of Sport Online* 7, no. 1 (2004): n. pag.

Longmore, Paul K. "Screening Stereotypes: Images of Disabled People in Television and Motion Pictures." In *Images of the Disabled, Disabling Images*, edited by A. Ian Gartner, 65–78. New York: Praeger, 1987.

Mitchell, David T., and Sharon L. Snyder. *Narrative Prosthesis: Disability and the Dependencies of Discourse.* Ann Arbor: University of Michigan Press, 2000.

Morris, Jenny. *Pride against Prejudice: Transforming Attitudes to Disability.* London: The Women's Press, 1996.

ProQuest. *British Periodicals.* Accessed July 5, 2018. https://www.proquest.com/products-services/british_periodicals.html.

Quayson, Ato. *Aesthetic Nervousness: Disability and the Crisis of Representation.* New York: Columbia University Press, 2007.

Roche, David. "The Metaphor of Facial Disfigurement." *Huffington Post*, May 25, 2011. http://www.huffingtonpost.com/david-roche/the-metaphor-of-facial-di_b_144949.html.

Silva, Carla Filomena, and P. David Howe. "The (In)validity of Supercrip Representation of Paralympian Athletes." *Journal of Sport and Social Issues* 36, no. 2 (2012): 174–94.

Wills, David. *Prosthesis.* Stanford: Stanford University Press, 1995.

Index[1]

[1] Note: Page numbers followed by 'n' refer to notes.

R. Sweet, *Prosthetic Body Parts in Nineteenth-Century Literature and Culture*, Palgrave Studies in Nineteenth-Century Writing and Culture, https://doi.org/10.1007/978-3-030-78589-5

The manufacturer's authorised representative in the EU is Springer Nature Customer Service Centre GmbH, Europaplatz 3, 69115 Heidelberg, Germany. If you have any concerns regarding our products, please contact ProductSafety@springernature.com

Printed and bound by CPI Group (UK) Ltd, Croydon, CR0 4YY
30/09/2026
02225459-0008